MODERNISATION STRATEGY FOR NATIONAL IRRIGATION SYSTEMS IN THE PHLIPPINES

BALANAC AND STA. MARIA RIVER IRRIGATION SYSTEMS

Mona Liza Fortunado Delos Reyes

Thesis committee

Promotor
Prof. Dr E. Schultz
Emeritus Professor of Land and Water Development
UNESCO-IHE Institute for Water Education
Delft, the Netherlands

Co-promotors
Prof. Guillermo Q. Tabios III, PhD, MSc
Prof. of the Institute of Civil Engineering
University of the Philippines Diliman
Manila, Philippines

Dr K. Prasad
Freelance Irrigation and Water Resources Management Consultant
Lalitpur, Nepal

Other members
Prof. Dr R. Uijlenhoet, Wageningen University & Research
Prof. Dr N.C. van de Giesen, Delft University of Technology
Prof. Dr Hector Malano, The University of Melbourne, Australia
Dr Robina Wahaj, Food and Agriculture Organisation of the United Nations, Lahore, Pakistan

This research was conducted under the auspices of the SENSE Research School for Socio-Economic and Natural Sciences of the Environment

MODERNISATION STRATEGY FOR NATIONAL IRRIGATION SYSTEMS IN THE PHILIPPINES

BALANAC AND STA. MARIA RIVER IRRIGATION SYSTEMS

Thesis

submitted in fulfilment of the requirements of
the Academic Board of Wageningen University and
the Academic Board of UNESCO-IHE Institute for Water Education
for the degree of doctor
to be defended in public
on Thursday 6 April, 2017 at 2:00 p.m.
in Delft, the Netherlands

By

Mona Liza Fortunado Delos Reyes
Born in Manila, Philippines

CRC Press/Balkema is an imprint of the Taylor & Francis Group, an informa business

Published by
CRC Press/Balkema
PO Box 11320, 2301 EH Leiden, The Netherlands
e-mail: Pub.NL@taylorandfrancis.com
www.crcpress.com – www.taylorandfrancis.com

www.crcpress.com - www.taylorandfrancis.co.uk - www.balkema.nl
ISBN 978-1-138-06774-5 (Taylor & Francis Group)
ISBN 978-94-6343-103-3 (Wageningen University)
DOI: http://dx.doi.org/10.18174/406141

TABLE OF CONTENTS

DEDICATION

This PhD thesis is dedicated in memory of the two educators whose aspirations served as an inspiration for this work: my father, Jesus M. Delos Reyes and my mentor, Prof. Wilfredo P. David.

ACKNOWLEDGEMENTS

A doctoral study is a challenging journey. It is both exciting and daunting at times. It is exciting because it gives one an opportunity to contribute in an effort to find just and effective ways for a greater good. An offshoot of such opportunity is just as stimulating - a chance to communicate and work with highly inspiring people in the academia, government agencies and in the field. Like in a journey, the roads are not always smooth. There can be bumps, mires, precarious turns and a seemingly interminable empty horizon.

The completion of this academic journey was made possible by guidance and goodwill of many individuals to whom I am deeply grateful.

I highly appreciate the guidance, support and encouragement of my promoter Em. Prof. E. Schultz. It is an honour to be your student. I am inspired by the wisdom that comes naturally from you.

I am indebted to my late supervisor Prof. Wilfredo P. David whose vast knowledge and deep understanding of the technical aspects and cross-cutting issues of my PhD thesis and whose wisdom and virtues in the practice of his profession were truly an inspiration. I was privileged to have learned from and worked with you.

I am grateful to my local supervisor, Prof. Guillermo Q. Tabios III for his invaluable guidance and encouragement particularly on the modelling aspects of this doctoral research. I have learned much from you. I am also thankful to Dr. Krishna Prasad for his valuable support and insights which helped this study reached milestones.

My gratitude also extends to dedicated officials and staff of the National Irrigation Administration (NIA) who willingly accommodated my request for data and shared their insights on various aspects of their mandates. I am thankful to officials and members of the irrigators associations of Sta. Maria RIS and Balanac RIS for their usual support and cooperation in the conduct of my various field experiments.

I am also indebted to the Philippine Government and the Netherlands Government for the financial support to my PhD study. I thank my host institution, UNESCO-IHE, for the vibrant and stimulating academic atmosphere. Also to Maria Romero Vink and family for generosity and kindness they have shown me during my stay in the Netherlands.

To my parents and siblings, for their presence in my life no matter what.

1. INTRODUCTION

1.1 General

Expanding the aggregate irrigation service area has been a key strategy of the Philippine Government to achieve its goals of rice self-sufficiency and alleviation of rural poverty. Construction of new irrigation facilities was vigorously carried out with the launch of a 10-year accelerated irrigation development program in 1974 (National Irrigation Administration (NIA), 1990). The irrigation service area increased from about 0.74 million ha in 1972 to about 1.44 million ha in 1985, almost 94%.

Rehabilitation of irrigation systems has also been carried out to maintain the irrigation service areas and to close the gap between the actual area irrigated and the design service area. It is defined by the NIA as an undertaking that includes reconstruction or restoration of irrigation facilities to their original design, expansion of the service areas beyond the original design and provision of additional control structures, drainage systems, on-farm facilities and service roads.

Since the late 1980s, however, irrigation expansion in the Philippines has lost momentum due to a host of interrelated technical, hydrologic, socio-economic, institutional, environmental and political issues and constraints (David, 2003). A number of studies revealed that many canal irrigation systems are performing below expectations (Ferguson 1987; Dy, 1990; World Bank, 1992; David, 2003; David, 2008; David 2009). The actual area irrigated is much less than the design service area especially during the dry season. Rice yield and cropping intensity are much below their projected levels. Consequently, post-evaluation economic rates of return of irrigation projects are much lower than those estimated during project planning. Further, the systems rapidly deteriorate due to cumulative operation and maintenance neglect and apparent too optimistic design assumptions that have compromised the physical integrity of these irrigation systems (David, 2003; 2008). Newer canal irrigation systems only served around 56% of their design service areas in contrast with up to 94% prior to 1965 as reported by Fergusson (1987).

The performance of large-scale canal irrigation systems remained generally less satisfactory even after the adoption of participatory irrigation management schemes

(Bagadion, 1994; Raby, 2000; David, 2003; International Programme on Technology and Research in Irrigation and Drainage (IPTRID), 2003; Maleza and Nishimura, 2007).

The policy of continuing public investment in the development and rehabilitation of canal irrigation systems has increasingly been questioned in the light of continued dismal performance of such irrigation systems, attractiveness of shallow tubewell irrigation, heavy foreign debt burden, pressing need to upgrade other rural infrastructure such as roads, markets, and post-harvest facilities. While the validity of such arguments cannot be denied, there is still a strong case for construction of new and modernisation of existing canal irrigation facilities from the standpoints of food security, poverty alleviation, equity in access to irrigation water, and preservation of the integrity of the water resources (David, 2003).

The dismal results of rehabilitation efforts have been attributed to the approach of restoring the original design without rectifying apparent unrealistic design considerations (David, 2003; David et al., 2012a; David et al., 2012b). IPTRID (2003) noted that the Philippine government borrowing for irrigation during the 1980s and 1990s has mostly supported rehabilitation and that maintenance was deferred. However, no assessment on the causes of rapid deterioration of flow control structures and no changes on water delivery were made.

Modernisation of irrigation is widely viewed as a process of technical, management and institutional transformation to improve irrigation services to farmers (Food and Agriculture Organisation of the United Nations (FAO), 1997; Renault et al., 2007). It is now recognized as a strategic option to improve the performance of large-scale canal irrigation systems. As each irrigation system is unique, a modernisation strategy must be system-specific.

In the Philippines, large-scale canal irrigation systems, which are more commonly called in the country as national irrigation systems (NIS), are designed based on rice crop water requirement and a water source supply with at least 80% probability of exceedance. On the average, however, many of these systems can irrigate only about 60–70% of their design service areas during the dry season despite the various rehabilitation efforts carried out by the government. This low percentage of area actually irrigated has been attributed to inadequate water supply or low irrigation efficiency or both.

This study discusses the proposed approach to investigate modernisation options for NIS. Central to the proposed modernisation strategy are:

- focus on establishing a coherent link among the physical structure, system operation and water supply in the analysis and formulation of a modernisation plan for NIS;

- consideration of adaptive modifications in the physical structure, operation, maintenance and management of irrigation systems in the context of changing hydrological regimes.

The available water supply would be the main limiting factor on how much of the service area can be irrigated. A prudent approach to planning for irrigation system modernisation would consider a realistic service area that can be irrigated with the available water supply. In many NIS, this would mean reducing their service areas. The advantages of reducing the service area down to the realistic level would include a more focused modernisation plan, a more reliable water service delivery, higher unit area productivity as a result of more reliable water supply and a higher chance of full operation and maintenance (O&M) cost recovery due to most likely increases in irrigation service fee collection as irrigation service to farmers is improved. Meanwhile, augmentation of the water supply would help maintain the design service areas. Conjunctive use of water from other sources and from the original water source for the irrigation system would help address inequity in access to irrigation water as well as improve reliability, flexibility and adequacy of irrigation service and, probably, simplify system operation.

The general methodology used in developing the approach included the following: analysis of the process, nature and impacts of rehabilitation projects; diagnostic assessment of the irrigation systems; revalidation of design assumptions on water balance parameters; characterization of irrigation management and demands; and identification of options for improvement.

1.2   Structure of the thesis

This thesis consists of eleven chapters. Chapter 1 highlights the trend and present state of irrigation development in the Philippines as well as the nature and scope of the study. Chapter 2 gives an overview of the state of irrigation in a global perspective; brief history of irrigation development in the Philippines, the present level of irrigation development and issues in the irrigation sector; and the rationale, research questions, hypothesis, objectives, and research methodology of the study. Chapter 3 provides descriptions of general hydro-climatic, socio-economic conditions, laws and policies of the country that are relevant to water resources and irrigation development. The description of the irrigation systems considered in this study is also described in this chapter. Chapter 4 presents a review of pertinent literature on the issues and concepts on design and performance of irrigation systems, while Chapter 5 summarizes the general approaches to assessment of irrigation systems.

Chapter 6 discusses the results of the assessment study on the impact of rehabilitation in increasing the service area actually irrigated. Chapter 7 presents the diagnostic assessment of the performance of the irrigation systems under study as well as the results of the analysis on consistency among design philosophy, operational objectives, physical structure and current irrigation water demand. The validity of design assumptions on percolation and water supply is investigated in Chapter 8. Chapter 9 deals with the water flow paths, system management, irrigation service and demand. Chapter 10 presents the options for improving the physical structure and system operation of the case study systems, visions of the modernised systems and the major steps in the formulation of a modernisation plan for NIS. The key findings for each objective and some of the challenges to NIS modernisation are discussed in Chapter 11. The thesis is concluded with an evaluation of the proposed approach to formulation of a modernisation plan for NIS and an indication of the way forward.

2. BACKGROUND AND OBJECTIVES

2.1 Irrigation in global perspective

The primary purpose of irrigation is the production of food and fibre. Agricultural production in areas far from water resources and during dry seasons has been made possible by extensive irrigation development. Irrigation development has been adopted as a key strategy of many countries to attain food security during the twentieth century (Turral et al., 2010). The increase in global food production during this period was closely associated with the expansion of irrigated land (Clemmens, 2006). Irrigation was a key element that underpinned the Green Revolution, the mid-20th century period was marked by an increased agricultural output in Asia (Hussain, 2007a; Turral et al., 2010). Irrigation offers a quick way of increasing food production through increased yields and cropping intensities (Abdullah, 2006). Approximately 40% of the world's crop production comes from the irrigated land, which is about 20% of the world's cultivated lands (FAO, 2007).

Irrigation has also been viewed as an instrument to increase rural household income and to alleviate rural poverty (Abdullah, 2006; Hussain, 2007a; Hussain, 2007b; Lipton, 2007; Molden et al., 2007; Turral et al., 2010). It generates employment opportunities and creates conditions for subsequent agro-industrial and economic development (IPTRID, 1999; Abdullah, 2006; Molden et al., 2007). Surplus in agricultural production made possible by irrigation is also an important source of foreign revenue for many countries (Dedrick et al., 2000).

By the period 2025–2030, global food production will have to be doubled to keep up with the food requirement of the concurrent world population (Schultz et al., 2005; Abdullah, 2006; Schultz et al., 2009) of around 8 billion people. It is expected that around 80–90% of this increase will have to come from existing cultivated areas and 10–20% from newly cultivated land (Schultz and Wrachien, 2002; Schultz et al., 2009). Since significant increase in yield cannot be expected from areas without irrigation facilities, around half of the required food production over the next couple of decades will to have to be realized on irrigated lands (Schultz and Wrachien, 2002; Schultz et al., 2005; Abdullah, 2006; Schultz et al., 2009).

However, not all is well with irrigation. Despite impressive contributions in global food production, there have been issues and concerns on the environmental impact, performance and sustainability of irrigated agriculture. Irrigation accounts for about 70% of water withdrawals from global river systems, while the irrigation efficiency of canal irrigation systems is generally less than 40% (Plusquellec et al., 1994). Excess water flowing out of irrigated areas has been associated to waterlogging, soil salinity, freshwater pollution by agro-chemicals and habitat for vectors of water related diseases (United Nations Conference on Environment and Development (UNCED), 1992).

It has also become a common knowledge that many canal irrigation systems have been performing poorly (Horst, 1998; Skutsch and Rydzewski, 2001; Plusquellec, 2002; Clemmens, 2006; Malano and Hofwegen, 2006). The economic rates of return, cropping intensities and actual area irrigated have been lower than expected during the project planning stage (Rice, 1997; Pluesquellec, 2002; David, 2003). Inability to achieve the performance targets in quite some canal irrigation systems has been attributed to the commonly observed problems, which include unreliable water supply, insufficient maintenance, poor system operation or water management, inappropriate flow control structures, poor water distribution to the detriment of downstream farmers, rapid deterioration of physical structures, and technical design shortcomings, among others.

The poor performance and the recurrent rehabilitation and operation and maintenance costs, which are hardly recovered from users, have casted doubts on the sustainability of most canal irrigation systems. Such lack of sustainability has been one of the key factors in the decrease of investments by international funding agencies in canal irrigation system development (Pluesquellec, 2002; Malano and Hofwegen, 2006).

A widespread installation of private tubewells for irrigation purposes, which started in the early 1980s (Turral et al., 2010), has been associated mainly to the unreliability of water delivery of canal irrigation systems (Plusquellec, 2002; IPTRID, 2003; Ertsen, 2009). Groundwater irrigation has been the dominant mode of irrigation in India and has been used extensively in China, Nepal, Bangladesh and Pakistan. As cited by Ertsen (2009), the total groundwater use in these countries accounts for 300 billion m^3yr^{-1} or nearly half of the estimated worldwide groundwater withdrawals of 600–700 billion m^3yr^{-1}. Groundwater irrigation is credited to have sustained the increase in global food production earlier achieved with a rapid development of canal irrigation systems. However, its success comes at a price: groundwater is being depleted at undesirable rates in many irrigated areas (Plusquellec, 2002; IPTRID, 2003; Ertsen, 2009). Continued reliance on groundwater irrigation in many countries will expedite reaching the limits on groundwater development.

There is a consensus that irrigation will play a key role in securing global food supply in the forthcoming decades. This role implies that the contribution of irrigated areas in food production will have to increase. Increasing such share can be achieved by expanding the areas with irrigation and drainage facilities and improving and modernizing existing irrigation and drainage systems (Schultz *et al.*, 2005; Schultz *et al.*, 2009; Turral *et al.*, 2010). The latter is the more subscribed option (Malano and Hofwegen, 2006; Schultz *et al.*, 2009; Turral *et al.*, 2010) as the former is deemed more costly and has limited potentials as competition for land and water among various uses (industries, cities, and environment) is expected to intensify. Consequently, performance improvement of existing canal irrigation systems has been the theme of many papers, international fora and development efforts on irrigation. Various performance assessment methodologies and performance indicators have been developed. Similarly, a number of concepts, processes and activities aimed at improving performance have been proposed. Shifting focus from mainly technical solutions pursued prior to 1980s to management and institutional solutions during the 1980s (Horst, 1998; Plusquellec, 2002; Malano and Hofwegen, 2006), many of them now gravitate towards a modernisation strategy that emphasizes consistency among design of physical components of irrigation systems, operational procedures and agreed water delivery service (rate, duration and frequency). Service-oriented approach to design and management of irrigation systems has recently been the dominating philosophy subscribed to by most irrigation experts. It is also the proposed guiding principle in modernisation efforts aimed at improving the performance of existing irrigation systems. Ensuant development includes redefining modernisation as a process of technical and managerial upgrading of irrigation systems to improve irrigation service to farmers. Such becoming commonly understood definition of modernisation distinguishes it from rehabilitation, which simply re-establishes the physical design of the original system (Plusquellec, 2002).

The service-oriented approach to irrigation development and modernisation presents a key challenge since a service-oriented design has to account for farmers demand and responses to water deliveries as a standard in irrigation. Thus, it would require knowledge on hydrology, hydraulics and social aspects (Ertsen, 2009). It implies a design process that specifies the physical components of an irrigation system within a well-defined operational procedure that enables delivery of an agreed level of water service (Plusquellec, 2002). It also calls for sufficient flexibility to be incorporated in the design to better respond to changes in water service demands (Schultz and De Wrachien, 2002). Further, it also necessitates a review or re-assessment of irrigation design procedures, criteria and assumptions (Plusquellec, 2002; Schultz and De Wrachien, 2002). Groundwater, which provides farmers water storage in the vicinity of their farms, presents opportunities to augment the surface water supply of canal irrigation systems. Conjunctive use of surface water and groundwater is deemed to provide a better irrigation service.

2.2   Irrigation in the Philippines

2.2.1   History of irrigation

Accounts on irrigation development in the Philippines can be found in the National Irrigation Administration (NIA) (1990) and David (2003). This review draws heavily from these references. The time when irrigation was first practised in the country predates recorded history. Historical accounts and archaeological findings approximate the age of the earliest irrigated rice terraces in the country at 2,000 years old. Similarly, early canal irrigation systems in lowland areas first existed prior to colonial times. These irrigation systems were constructed and operated by local farmers. Water for these communal irrigation systems was usually diverted from a river by placing bamboo and rock structures across the river. Based on Spanish reports Zanjera, an indigenous irrigation society found mostly in the Ilocos Region, came into existence starting 1630 (NIA, 1990).

Prior to the Spanish colonial era, it was estimated that an aggregate of about 25,000 ha in the country was served by mainly very small canal irrigation systems, which were built, operated and maintained by farmers. David (2003) remarked that an interesting feature of these systems was their built-in stability as a result of crop cultural practices, cropping patterns and soil and water conservation and management practices being interwoven in their design, operation and maintenance.

The Spanish colonial government constructed irrigation systems in lands granted by its King to Spanish religious orders in the country. These religious orders grew rice as a major crop in these lands, which was familiarly known as "friar lands." The colonial government introduced new techniques and designs in the construction of the irrigation systems serving friar lands in provinces around Manila. Typical to these irrigation systems are massive headworks up to 40 metres high and large networks of canals to convey water (NIA, 1990).

During the Spanish regime, many small run-of-the-rifer irrigation systems were built as well to serve mainly friar lands in coastal areas close to Manila, along the Pampanga River and in the Ilocos Region. These irrigation schemes served an aggregate area of 200,000 ha (David, 2003).

During the American regime (1900–1936), the Irrigation Division under the Bureau of Public Works was established in 1908 and the Irrigation Act was passed in 1912. The Irrigation Division was responsible for constructing new irrigation facilities and repairing existing ones. The Irrigation Act regulated the appropriation of public waters, prescribed rules on water rights, and provided for the investigation, construction, operation and maintenance of irrigation systems and payments thereof. David (2003) noted that this Act was significant because it attempted to integrate the planning, design, construction and operation and maintenance (O&M) of irrigation systems under one institution. It transferred the operational control of irrigation facilities from the regional offices of the Bureau of Lands to the Irrigation Division. However, he argued that it paved the way for the centralization of irrigation development and set in motion a process of viewing irrigation development projects mainly from technical and engineering perspective, with focus on physical infrastructure.

The Irrigation Division constructed the first national irrigation system (NIS) in the country, the San Miguel River Irrigation System in Tarlac Province, with a service area of 6,000 ha. This NIS was inaugurated in 1913. From 1922 to 1930, 11 national irrigation systems with a total irrigated area of around 80,000 ha were completed (NIA, 1990). While the emphasis was on small canal irrigation systems for rice monocultures, 12 medium-sized national irrigation systems with a total service area of 91,000 ha of rice farms were constructed during the American regime (David, 2003).

During the Commonwealth period, government funds for communal irrigation projects were coursed through the legislators. With its evident political purpose, these funds were usually spread out thinly over many public works projects within the legislator's district. This resulted in the construction of dams across streams with insufficient water supply or on sites where foundations were unstable and, in many instances, unfinished irrigation projects (NIA, 1990).

During the Japanese regime, only two national irrigation systems with total service area of about 1,270 ha were constructed (NIA, 1990). Irrigation development activity during this period was minimal as safety and survival during the war against the Japanese invasion was the overriding concern of the nation. At the end of this period, many irrigation systems were in a bad state of deterioration and disrepair (David, 2003).

The initial efforts on irrigation during the early independence (1947–1965) were aimed at rehabilitating existing systems neglected or damaged during the World War II. Succeeding activities increased the total number of NIS to 44 with an aggregate service area of 196,650 ha. Including areas irrigated by other public-funded systems, the total irrigated area in the country at the end of 1957 was about 400,000 ha.

Pump irrigation development as government program started during the early independence with the establishment of the Irrigation Pump Administration (IRPA) under the then Department of Agriculture and Natural Resources in 1949. The task of IRPA included the purchase of pump irrigation equipment and supervision of their installation, operation and maintenance (NIA, 1990). In 1952, IRPA was reorganized as the Irrigation Service Unit (ISU). David (2003) stated that the establishment of the ISU was significant in two respects. First it placed under one institution the planning and implementation of certain development activities and the delivery of essential irrigation support services and functions. Second, it gave individual farmers access to an easy-to-operate irrigation system and, hence, greater control over its O&M.

Prompted by the need to construct more irrigation systems, a bill was signed into law (also known as Republic Act No. 3601) that created the National Irrigation Administration (NIA) in 1963. The implementation of this Act was made effective in 1964. All functions, assets and liabilities of the abolished Bureau of Public Works Irrigation Division were absorbed by NIA. The Irrigation Unit of the Bureau of Lands and the Friars Lands Irrigation

Systems were also transferred to NIA. The NIA has become a public corporation primarily responsible for the irrigation development. In 1964, the country had 79 NIS, 771 communal irrigation systems and 2,540 pump irrigation systems with a total service area of 541,000 ha.

At the latter part of the period of early independence, the focus of irrigation development shifted towards large-scale multipurpose development projects (David, 2003). Among the first of these was the Upper Pampanga River Irrigation System (UPRIS), whose planning and feasibility study was undertaken by the United States Bureau of Reclamation (USBR) in pursuant to the National Economic Council and the United States Agency for International Development (NEC-USAID) agreement in 1962 to establish a planning program for the water resources development in seven major river basins in the country.

A rapid expansion of rice area with irrigation facilities occurred between 1966 and 1988. After the UPRIS feasibility study in 1966, the USBR also submitted a feasibility report for proposed projects on other six major river basins between November 1966 and January 1967. Construction of new irrigation facilities was vigorously carried out with the launch of a 10-year accelerated irrigation development program in 1974 (NIA, 1990). The irrigation service area increased from about 0.74 million ha in 1972 to about 1.4 million ha in 1985, almost 94%. In irrigation development more emphasis was put on medium- and large-scale canal irrigation systems. It was heavily subsidized by the government.

During this period of rapid expansion in irrigation, other government agencies became involved in irrigation development: the Farm Systems Development Corporation (FSDC) and the Bureau of Soil and Water Management (BSWM). The former was mandated to develop small (less than 1,000 ha) low-lift pump irrigation schemes while the latter started developing small water impounding irrigation systems. Before the end of the period, FSDC was abolished and NIA stopped its pump dispersal program.

In the late 1980s, performances of irrigation systems way below expectations during the planning stage hinted that not all was well in irrigation development in the country. David (2003) pointed that the policy bias towards large canal irrigation infrastructure was not conducive to private sector participation in irrigation development and judicious choice of more suitable and efficient irrigation technologies by farmers. The public investment per unit irrigation service area for large canal irrigation systems has steadily increased while some 30–50% of the service area was not being irrigated and physical infrastructures were rapidly deteriorating despite of rehabilitation efforts.

The policy of continuing to invest in the development of new and rehabilitation of medium to large-scale canal irrigation systems has increasingly been questioned in the light of sustainability issues and attractiveness of other modes of irrigation. A rationalization of irrigation development has started in the late 1980s and has resulted in: (1) the transfer of NIA to the Department of Agriculture; (2) the search for more-cost effective, efficient and sustainable irrigation technologies; and (3) policy reforms and institutional changes to promote such technologies (David, 2003). The transfer of NIA to the Department of Agriculture was deemed as an appropriate step to have a better coordination of irrigation development activities with the delivery of essential irrigated agriculture support services. Similarly, the launching in 1992 by the Department of Agriculture of a shallow tubewell (STW) irrigation project was viewed as a big step towards increasing farmers' participation in irrigation development and their control over irrigation facilities and gradually removing irrigation subsidy. Shallow tubewells have become popular with farmers as they can exercise greater degree of control over water and their cropping system. In the late 1990s, the shallow tubewells have increased at a rate of over 10%. Over 90% of all existing shallow tubewells were installed purely through the initiatives of the private sector.

In the late 1990s, the modernisation of the agriculture sector was accorded top priority by the government in an effort to provide enduring solutions to the twin problem of food insecurity and rural poverty. The primary instrument for doing this was to expand the irrigation base in an efficient and cost-effective manner. The Philippine Congress passed the Agriculture and Fisheries Modernisation Act (AFMA) in 1997. The AFMA stipulated an agenda for action to rationalize irrigation development in the country. The agenda called for development of additional irrigation command areas; effective and efficient rehabilitation, operation and maintenance of existing irrigation systems; and enhanced institutional capacity for the effective and timely delivery of essential irrigated agricultural services and functions guided by sound policies and programs. These policies and programs emphasize expenditure switching for cost-effectiveness and improved performance of irrigation systems, expanding the technology base for rectifying design shortcomings, improving the operation and maintenance of irrigation systems and increasing the productivity of irrigation water, institutional reforms and strengthening; genuine devolution of activities, services and functions to local government units and the private sector, and a system of accountability in the expenditures of public funds for irrigation. Section 28 of the Act stipulates the selection of appropriate irrigation schemes for development or rehabilitation shall be based on the following criteria: (1) technical feasibility; (2) cost-effectiveness; (3) affordability or low investment cost per unit area; (4) sustainability and simplicity of operation; (5) O&M cost recovery; (6) efficiency in water use; (7) length of gestation period; and (8) potential for increasing unit area productivity. David (2006; 2009) emphasized that these criteria are not mutually exclusive and that the first four are the minimum criteria, hence, must all be met.

In recent years, rehabilitation has also been viewed as a requisite to a successful transfer of the Government-managed national irrigation systems (NIS) to irrigators associations (IA). With the rehabilitated physical structures, it is expected that the IAs would be willing and able to take over the operation, maintenance and management of the NIS. Individual experts within the NIA have recognized the mediocre impact of past rehabilitation efforts and advocated a modernisation-based irrigation rehabilitation reform. Modernisation-based irrigation rehabilitation is aimed at improving diversion and conveyance capacities, water control and allocation, and water adequacy and productivity of irrigation systems. It is envisioned to transfer the function of water distribution from irrigation stewards to irrigation structures.

2.2.2 Present level of irrigation development

The NIA (2014) estimated the total irrigable area of the country to be about 3 million ha. About 1.78 million ha or 57% of this irrigable area are provided with irrigation facilities (Table 2.1 and Figure 2.1). Of these areas equipped with irrigation structures, 46, 43 and 11% are service areas of national irrigation systems (NIS), communal irrigation systems (CIS) and private irrigation systems, respectively. By end of 2014, the area under irrigation totalled 1.71 million ha (Table 2.2). The reduction in the total service area accounted for the service areas converted to non-agricultural use and non-restorable areas that were previously developed for irrigation.

NIS are publicly funded and government-owned systems. They can be co-managed with water users associations that have signed a contract for irrigation management transfer (IMT). CIS are owned and operated by group of farmers while private irrigation systems are mostly individually-owned small pump irrigation. As of December 2014, there are 245 NIS and 10,651 CIS as well as 2,485 water users associations that have entered into IMT contracts. All NIS and CIS are canal, gravity type irrigation systems.

Table 2.1. Service areas of the different irrigation schemes in selected years

Year	Cumulative Developed Service Area, ha			
	NIS	CIS	Pump	Total
1980	472,000	577,000	152,000	1,201,000
1985	568,000	704,000	152,000	1,424,000
1990	663,209	750,671	152,128	1,566,008
1995	651,812	474,289	180,909	1,307,010
2000	685,812	501,442	174,200	1,361,454
2005	695,774	543,262	174,200	1,413,236
2010	767,006	558,333	217,329	1,542,668
2014	821,598	765,873[a]	194,841	1,782,312

[a] Included other government assisted CISs

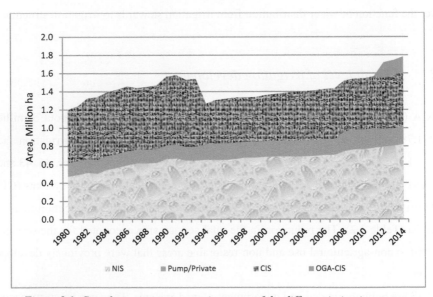

Figure 2.1. Cumulative irrigation service areas of the different irrigation systems

Table 2.2. Status of irrigation development in hectares

Region	Irrigable area[a]	Firmed up service area[b]				Total	Irrigation development, %
		NIS	CIS	OGA-CIS	Private		
CAR	97,310	13,996	46,991	1,667	27,073	89,727	92
1	262,744	46,684	51,014	50,722	21,240	169,659	65
2	456,898	149,282	53,396	23,81	49,499	275,987	60
3	480,783	194,369	67,312	22,357	7,792	291,830	61
4A	85,929	20,552	18,425	2,578	6,334	47,889	56
4B	138,719	18,938	31,780	4,201	14,469	69,387	50
5	239,440	23,189	71,563	13,035	25,059	132,846	55
6	189,934	47,090	38,257	15,459	15,053	115,858	61
7	46,159	11,538	25,144	1,371	4,718	42,771	93
8	84,081	21,348	38,030	3,286	6,197	68,861	82
9	74,952	16,361	23,583	3,509	1,817	45,270	60
10	113,631	25,827	25,213	3,659	6,254	60,953	54
11	147,313	35,466	24,327	2,812	1,636	64,241	44
12	286,263	66,377	35,671	10,836	3,315	116,199	41
ARMM	156,205	25,643	19,539	295	90	45,567	29
CARAGA	159,249	33,510	25,409	7,802	4,297	71,018	45
Total	3,019,609	750,169	595,653	167,400	194,841	1,708,063	57

Source: NIA, 2014. [a] Based on rice and corn areas with 3% slope. [b] OGA - other government-assisted CIS

About 60% of the total NIS have run-off-the-river diversion dams (Table 2.3). Almost 40% the total NIS have service areas less than 1,000 ha and are classified as small-scale irrigation systems. Only about 13% has firmed-up service area (FUSA) greater than 5,000 ha. The two largest NIS are the Upper Pampanga River Integrated Irrigation System (UPRIIS) and Magat River Integrated Irrigation System (MARIIS), which are also used for flood control and power generation. Each is consisted of a cluster of irrigation systems. The main dam of UPRIIS (Pantabangan Dam) is 107 m high, 1,615 m and together with other smaller systems, serves 120,000 ha. MARIIS serves an aggregate area of 85,000 ha. Its main dam (Magat Dam) 114 m high and 4,160 m long.

Table 2.3 Categories for national irrigation systems

By FUSA, ha	
1,000	95
1,001- 5,000	118
5,001 - 10,000	15
10,001 - 20,000	14
20,001 - 50,000	2
50,000 - 65,000	1
By type of headwork	
Runoff-the-River diversion dam	149
Reservoir-type dam	13
Intake barrel-type diversion	66
Pump-type diversion	18

Data source: MID-NIA (2015)

2.2.3   Other irrigation related issues

David (2003) presented a number of issues and constraints to sustainable irrigation development in the country, which included, among others, the absence of significant irrigation research and development (R&D) efforts and the fragmentation of irrigation development activities. There were little useful R&D outputs for the formulation of irrigation development plans and programs, establishment of design criteria and development of appropriate irrigation technologies. He pointed out the need for a comprehensive water resources policy framework that will take into consideration the competing, multiple uses of water and lend coherence to existing laws, policy guidelines and regulations on water rights, water pricing and water control. He stated a number of requisites for a national water commission that should be established for the task of developing such water resources policy framework. The National Economic Development Authority (NEDA) (2012) reaffirmed the need to address institutional fragmentation and lack of leadership in the water sector, dearth of reliable data, absence of scientific decision support systems and inadequate provisions of water-related infrastructures. The Government is working for the creation of an apex, central agency that will govern the development and management of the country's water resources.

Irrigation is next to domestic and municipal uses in the priority appropriation pursuant to the provisions of the Water Code of the Philippines. With increasing water demands for domestic, municipal, industrial, power generation and other uses, the irrigation sector faces uncertainty in meeting its water needs, especially during periods of prolonged dry seasons.

Araral (2005) argued that the problem of poor performances of irrigation systems in the Philippines is linked to inherent problems in incentive structure of a public bureaucracy like the NIA. The incentive problems include the valuation of agency performance and outputs, limited competition and employment rules and inflexibility of civil service system. In the case of the NIA, these are manifested in an oversized and less efficient bureaucracy, the promotion of farmers' participation with patronage and a tendency to build-neglect-build irrigation systems. Araral (2005) opined that these incentive problems were firmly established by the path-dependent irrigation development and reinforced by moral hazard and fungible nature of irrigation aid. He further argued that the combination of inherent bureaucratic incentives and easy access to foreign aids have not encouraged the NIA to vigorously undergo essential governance reforms.

The landmark legislative effort to modernize the country's irrigated agriculture has lost momentum with the non-implementation of AFMA. AFMA was not implemented mainly because of the lack of budget allocations to carry out its programs (Dy *et al.*, 2006). Consequently, its objectives are yet to be achieved and hardly made a dent in addressing the issues in irrigation development and arresting further deterioration of irrigation systems in the country (David, 2008, 2009).

The implementation of the NIA Rationalization Plan reduced the number of regular staff by 40%, with more than 2,100 senior field personnel, voluntary or otherwise, availing offers of early retirement (NIA, 2012). A number of individual experts opined that such reduction in technical field personnel would negatively impact on the technical capacity of the NIA to carry out its irrigation development mandate in the immediate years following the retrenchment of personnel. When the PhP[1] 11 billion NIA budget was increased by at least 200% starting 2011, NIA faced the dilemma of increased expected physical accomplishments in irrigation projects with reduced expertise and less experienced technical personnel.

2.3 Description of the research problem

The viability and effectiveness of an irrigation system depend on the soundness of the criteria used in its design. The design shortcomings of an irrigation system manifest themselves in a low ratio of actual areas irrigated to the design service area, a faster rate of deterioration of the structures as compared to the design economic life and unwieldy problems in operation and maintenance of the system (David, 2003).

Faulty designs often stem from inaccurate or inappropriate design criteria and assumptions. Cases of inappropriate irrigation design criteria are prevalent in the country because of the practice of generalizing the design criteria or parameters over regions or even larger areas.

Design is usually carried out in isolation from systems operation and maintenance. Design engineers are not required to test run, in collaboration with those who are supposed to operate and maintain them, the systems they designed. As a result, design mistakes are repeated from system to system without being rectified. Hence, there is a compelling need to re-evaluate the soundness or suitability of the design criteria used in the planning and development of gravity irrigation systems and to develop appropriate location-specific irrigation designs for improved performance of irrigation systems.

Meanwhile, modifications in the river basin (deforestation) and impacts of apparent change in weather patterns are projected to further increase the risks to irrigated crop production. Deforestation causes shifting and alteration of hydrographs. Dry season flows of

---

[1] PhP = Philippine Peso (1 PhP = 0.022 US$, price level 2015)

many rivers continue to decrease. More pronounced dry seasons, stronger typhoons, and bigger-magnitude, short-duration floods are being observed more frequently than usual. Shifting the bulk of food and agricultural production to the typhoon-free season (dry season) by insuring water supply and improving on-farm water control and management through improved system design is one pragmatic approach of sustaining agricultural growth and adapting to apparent change in weather patterns. Ensuring irrigation water supply during the dry season in areas presently served by canal irrigation systems would require additional sources of irrigation water. Groundwater provides storage of water which can be tapped for supplemental irrigation. However, it has not been considered in the conventional designs of most canal irrigation systems. Since most of the irrigation service areas of existing irrigation systems are underlain by shallow aquifers, groundwater utilization for supplemental irrigation would make a relevant design criterion in the development of modernisation measures for existing canal irrigation systems.

Private groundwater irrigation has become popular among farmers mainly because it gives higher reliability of water supply as the source is just in the vicinity of the farms and greater farmer's control in planning their cropping and irrigation strategies. It emphasizes the possibility of decoupling canal operation, especially for higher level canals, and irrigation at the farm level, especially the downstream farms. Such possibility may open opportunities to simplify operational procedures and present options for changes in flow control structures along higher level canals (Ertsen, 2009). Therefore, ways to best integrate the concept of conjunctive use of groundwater and surface water in the design of service-oriented irrigation systems have to be explored.

In recent years, modernisation has been viewed as a process of technical and managerial upgrading to improve irrigation service to farmers. This becoming widely accepted definition of modernisation distinguishes it from the conventional rehabilitation, which simply restore the physical design of the original system (Plusquellec, 2002). The Food and Agriculture Organization of the United Nations (FAO) put emphasis on modernisation as a process of improving the ability of the irrigation system to serve the current demands of farmers as well as to provide flexibility to respond to future needs with the best use of the available resources and technologies (Renault *et al.*, 2007). This service-oriented approach to system improvement and management is also espoused by many experts (Plusquellec, 2002;

Clemmens, 2006; Malano and Hofwegen, 2006; Molden *et al.*, 2007; Renault *et al.*, 2007; Ersten, 2009).

It is apparent that interventions that would substantially improve irrigation system performance go beyond the current rehabilitation approach. They will have to address design shortcomings and changes in water supply and irrigation demands as well as take stock of the resources and potentials of an irrigation system. Given the multi-faceted nature of irrigation, crafting a service-oriented irrigation modernisation plan is arguably a difficult task. However, the need to pursue irrigation improvement in a more efficient and cost-effective way has been compelling. Consequently, a strategy for appropriate modernisation plan has to be formulated.

Based on current general knowledge in irrigation and existing research tools, this study has conceptualized a strategy on formulating a modernisation plan for canal irrigation systems in the Philippines. Elements of the modernisation strategy specifically investigated are: (1) coherence of design, operation and water supply; (2)validity of design assumptions on percolation and water supply; (3) water management in the context of changing hydrological regime; and (4) the use of groundwater in conjunction with surface water in part of service areas experiencing water shortages. The approach to developing a modernisation plan has been tested in two national irrigation systems, namely: Balanac River Irrigation System and Sta Maria River Irrigation System.

2.4 Research questions

The following specific research questions will be answered by the research:
- what is the nature of rehabilitation methods or modernisation approaches being followed for large-scale canal irrigation systems?
- what is the level of consistency among design philosophy, operational objectives, distribution network and flow control structures and operational reality to meet the preset water delivery targets?
- how do assumptions on water balance parameters differ from location-specific data?
- to what extent are the estimates of water demand and irrigation supply based on the assumed water balance parameters influence the ratio of actual area irrigated to the design service area?

- how much area can be irrigated with the given water supply and existing and preferred cropping patterns?
- what potentials do exist in various NIS for supplementing their surface water supply with groundwater from confined shallow aquifers?
- what are options of using water from these aquifers in conjunction with the NIS surface irrigation (that is, where, when, how shallow tubewell irrigation be used conjunctively)?
- what design guidelines can be adopted in modernizing NIS amenable to conjunctive water use strategy?

2.5 Research hypothesis

Despite the Philippine government's massive investment in rehabilitation and advocacy on participatory irrigation management, the performance of publicly funded large-scale, agency-managed canal irrigation systems continues to be mediocre. Rehabilitation approaches are focused on restoring and improving the original physical structure and are not adequately addressing the root causes of the low performance of irrigation systems, which are: design philosophy and design criteria based on unrealistic assumptions on water balance parameters; flawed conceptualization of canal hydraulics; and ensuing choices of irrigation technologies that are in conflict with operational reality. The hypothesis of the study is hence formulated as follows:

Effective modernisation of national irrigation systems (NIS) in the Philippines can be achieved by rectifying past irrigation design shortcomings and adopting an appropriate design philosophy and design criteria, including those for conjunctive use of surface water and groundwater.

2.6 Research objectives

The general objective of this research is to identify technically feasible and hydrologically sound modernisation options for existing government-owned national irrigation systems (NIS) in the Philippines. The specific objectives of the research include the following:

- to analyse the nature of previous rehabilitation or modernisation efforts and their impacts in terms of closing the gap between the actual area irrigated and the design and firmed-up service areas;

- to assess the efficacy of the physical structure and operations of irrigation systems to deliver the design rate, duration and frequency of irrigation water or a modified irrigation water demand at the offtakes of secondary canals;

- to assess the validity of design values for dependable water supply, crop water requirement, seepage and percolation;

- to identify modernisation options for the irrigated NIS service areas;

- to investigate the potentials of shallow tubewell irrigation in the part of NIS service areas that cannot be fully irrigated with surface water as designed;

- to identify possible options of integrating conjunctive use of groundwater and surface water in the formulation of modernisation plans for NIS.

2.7　General research methodology

A number of guidelines for modernisation of NIS were proposed, namely: reduction in the service area of a NIS to be modernized down to realistic levels and irrigating the parts of service area that are usually short of water by a more efficient alternative mode of irrigation; full O&M cost recovery of the modernized system; greater degree of farmers' control over irrigation water; validation and improvement of the design philosophy and criteria; and rectification of design shortcomings prior to modernisation (David, 2003; 2008). This research builds on these guidelines and is aimed at developing a strategy for technically and hydrologically sound modernisation of NIS. It has used the rapid appraisal survey (RAP), Mapping System and Services for Canal Operation Techniques (MASSCOTE) approach, system design and improvement analyses and field measurements of design values as research tools:

The conceptual framework of the research methodology is shown in Figure 2.2.

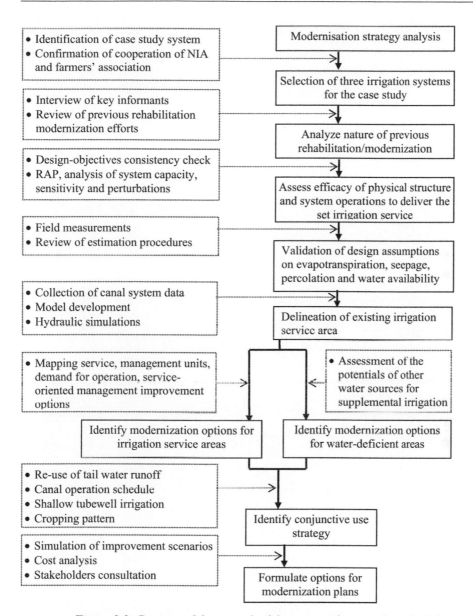

Figure 2.2. Conceptual framework of the proposed research methodology

3. DESCRIPTION OF THE PHILIPPINES AND IRRIGATION SYSTEMS OF THE CASE STUDY

3.1 Brief hydro-geographic description and general climatic conditions

The Philippines is located between 116° 40' and 126° 34' E and 4° 40' and 21° 10' N and is bordered by the Philippine Sea to the East, the West Philippine Sea (South China Sea) to the West and by the Celebes Sea to the South (Figure 3.1). It is an archipelago consisting of about 7,107 islands with a total land area of approximately 300,000 km². Its coastline stretches up to 36,289 km and is the fifth longest in the world.

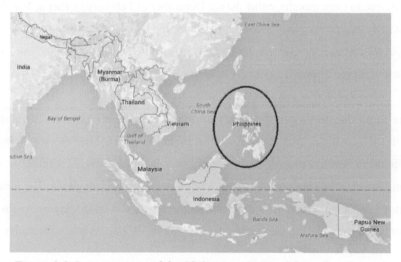

Figure 3.1. Location map of the Philippines (snipped from Google map)

The country is situated on the western fringes of the Pacific Ring of Fire. Most of its mountainous islands are of volcanic origin and covered by rainforests. The highest mountain (Mt Apo) measures up to 2,954 m+MSL (mean sea level), while the deepest point in the country (Galathea Depth in the Philippine Trench) is third deepest in the world.

The Philippine has a tropical and maritime climate, which is characterized by relatively high humidity, high temperature and abundant rainfall. The country's weather and climate are greatly influenced by temperature, humidity and rainfall.

Rainfall is the most important climatic element in the country. The rainfall distribution varies throughout the archipelago depending on the direction of winds and presence of mountain ranges. The mean annual rainfall ranges from 965 to 4,084 mm. Based on the distribution of rainfall, areas in the country are classified under four climate types (Figure 3.2) which are described as follows:

- *Type 1:* There are two pronounced seasons, dry from November to April and wet for the rest of the year. Maximum dry period is from June to September.
- *Type 2:* There is no dry season with a very pronounced maximum rain period from December to February. There is not a single dry month. Minimum rainfall occurs during March to May.
- *Type 3:* There is no very pronounced maximum rain period with dry season lasting only from one to three months, either during December to February or during March to May. This type resembles type 1 since it has a short dry season.
- *Type 4:* Rainfall is more or less evenly distributed throughout the year. This type resembles type 2 since it has no dry season.

The mean annual temperature based on the average of all but one (Baguio) weather stations in the country is 26.6 °C. The coolest month is January and the warmest month is May with mean temperatures of 25.5 °C and 28.3 °C, respectively. Altitude is the more significant factor in the variation of temperature than latitude and longitude. The mean annual temperatures of the northernmost station and southernmost station in the country do not vary significantly. Meanwhile the mean annual temperature of Baguio at an elevation of 1,500 m+MSL is 18.3 °C.

The Philippines has a high relative humidity owing its high temperature and surrounding bodies of water. The average monthly relative humidity ranges from 71% in March to 85% in September. Temperature and humidity reach their maximum levels during March to May, giving rise to high sensible temperature throughout the archipelago.

The climate of the country is defined by rainfall and temperature and has two major seasons: the rainy season from June to November and the dry season from December to May. The latter is subdivided into the cool dry season from December to February and the hot dry season from March to May.

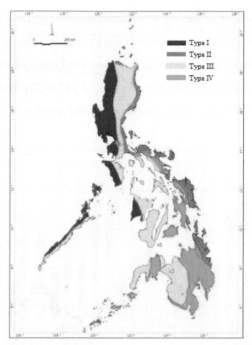

Figure 3.2. Climate map of the Philippines based on Corona's classification (adopted from Philippine Atmospheric, Geophysical and Astronomical Services Administration (PAGASA) http://www.pagasa.dost.gov.ph/index.php/climate-of-the-philippines)

Typhoons and El Niño are the two major weather disturbances in the Philippines. The country is at high risk of typhoons or intense tropical cyclones as it is situated in the busiest typhoon belt of the world - Western North Pacific, where the highest percentage (30%) of typhoons is formed. The combination of factors that favours the formation of most of the typhoons in the Philippines Sea and West Pacific Ocean include the following: (1) warm sea surface temperature (at least 26.5 °C with a depth of 50 m (150 ft)) and high humidity; (2) presence of intertropical convergence zone (ITCZ); (3) existence of low pressure areas within the ITCZ; and (4) light winds (less than 36 km/hr (23 mph)) in the upper atmosphere (6,000-15,000 m+MSL (20-50,000 ft+MSL)).

On the average, about 18 tropical cyclones per year enter the Philippine area of responsibility (Figure 3.3). Tropical cyclones are classified in the country according to their maximum sustained wind as follows:

- *Tropical depression:* with maximum sustained winds of up to 61 kph;
- *Tropical storm:* with maximum sustained wind speed of 62 to 88 kph;
- *Severe tropical storm:* with maximum sustained wind speed of 89 to 117 kph;
- *Typhoon:* with maximum sustained wind speed of 118 to 220 kph;
- *Super typhoon:* with wind maximum sustained wind speed exceeding 220 kph.

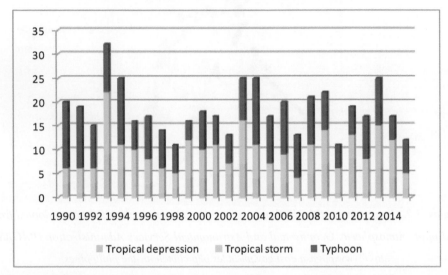

Figure 3.3.Frequency of tropical cyclones occurrence in the Philippines (data source: Philippine Statistics Authority (PSA, 2014))

The country is also prone to El Niño, the periodic, above-average warming of sea surface temperatures (SST) in the central and eastern tropical Pacific Ocean. The effect of an El Niño event in the country is usually a prolonged period of subnormal rainfall, which is accompanied by variations on distribution patterns of runoff, other hydrologic sub-processes and tropical cyclones. The effect of El Niño on rainfall and tropical cyclones depends on its intensity, area of coverage, duration and time of occurrence (De los Reyes and David, 2006). Strong El Niño events reduced monthly rainfalls by more than 50% during their peak period and suppressed tropical cyclone activity. Weak-to-moderate El Niño events did not reduce the occurrence of tropical cyclones, but shifted their tracks northeast.

3.2 Economy and agriculture

The Philippine economy is the 39th largest in the world (International Monetary Fund (IMF), 2014) and is one of the emerging markets. Its annual growth rates for the period 2010-2015 averaged 6.2% (Figure 3.4). It was driven by the services and industry sectors (Figure 3.5). Meanwhile, the agricultural sector had registered either contraction or minimal gain. The transport, storage and communication (TSC); trade; real estate, renting and business activities subsectors were major contributors to the growth of the services sector. Communication contributed robustly to the growth of TSC. The country has become the world's largest centre for business process outsourcing since recent years. Manufacturing and construction were the biggest contributors to the growth of the industry sector. The growth of agriculture for 2015 was pulled down by poor performance of rice, fisheries, corn, sugarcane and rubber.

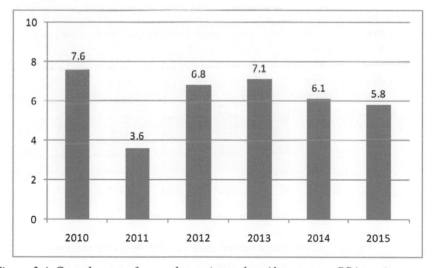

Figure 3.4. Growth rates of gross domestic product (data source: PSA, various years)

Though the agricultural sector accounted for only about 10% of the gross domestic product (GDP) (Figure 3.6), it remains large as it employed about 30% of the total labour force in the country (Figure 3.7). These figures would indicate low productivity in agriculture, that is, 30% of the labour force had to share in only 10% of the GDP.

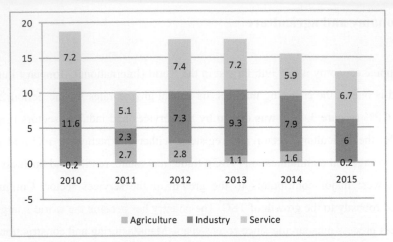

Figure 3.5. Growth rates of gross domestic product by sector origins

(data source: PSA, various years)

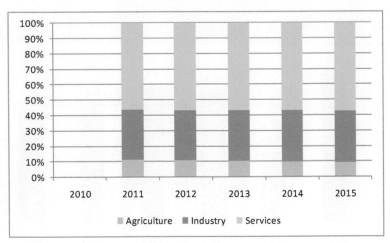

Figure 3.6.Percentage contribution to GDP by sector origins

(data source: PSA, various years)

The low productivity in agriculture is viewed as one of the main causes of high poverty incidence in rural areas of the country (Asian Development Bank (ADB), 2009; Ravago and Balisacan, 2015). Poverty in the country is both a rural and agricultural phenomenon. About 75% of the poor are in rural areas and about two-thirds are in the agriculture sector (Table 3.1). The highest incidence of poverty was among fishermen and farmers with about 40% and

38%, respectively. The agriculture sector is crucial for inclusive growth as agriculture and other agriculture-related employments remains the major source of income in rural areas where most of the country's poor live.

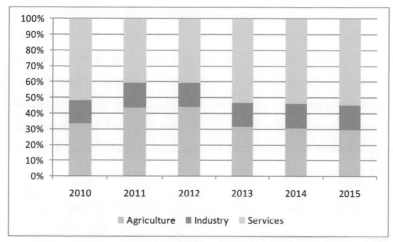

Figure 3.7.Percentage distribution of employment by sectors
(data source: PSA, various years)

Table 3.1. Magnitude of poor population and poverty incidence

	2006	2009	2012
Magnitude (in million)			
Total poor population	22.64	23.30	23.75
Urban dwellers	5.31	5.71	-
Farmers	1.77	1.69	-
Fishermen	0.40	0.35	-
Poverty incidence (%)			
Total population	26.6	26.3	25.2
Urban poor	12.6	12.6	13.0
Farmers	38.5	38.0	38.3
Fishermen	41.2	41.3	39.2

Data source: National Statistical Coordination Board (NSCB), 2009, 2013

- No official data released for 2012

The recurrence of natural extreme events and disasters has been among the major challenges in an effort to sustain agricultural growth. The country experienced an annual average of 201 natural disasters during the 2003-2013 period (PSA, 2014). The number of occurrence ranged from 66 events in 2005 to 374 events in 2011. Almost 94% (PhP 288 billion) of the economic losses were attributable to the direct and indirect effects of meteorological disasters, which were associated with tropical cyclones, tornados, continuous heavy rains, big waves, Southwest monsoon and cold front, among others (Figure 3.8). Destructive tropical cyclones caused the most economic damage costing PhP 278 billion or about 90% of the total economic losses during the 11-year period. In contrast, only about 4% and 2% of the economic losses were due to climatological and hydrological disasters. Climatological disasters were mainly El Niño-induced drought, while hydrological include flash floods or flooding and storm surge.

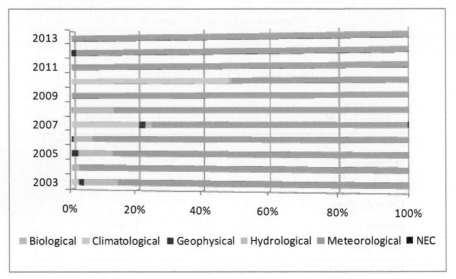

Figure 3.8.Percentage distribution of economic loss due to natural extreme events and disasters, by type of disasters (data source: PSA, 2014)

Agriculture sector suffered the most losses during the 11-year period with PhP 148 billion costs of damages or about 48% of the total economic losses (Figure 3.9). These were the biggest losses except in 2009 and 2013, when the biggest costs of damage were incurred in infrastructures and in private and communication, respectively. The losses in infrastructures totalled PhP 97.2 billion or about 32% of the total economic loss. The cost of damages in private and communication amounted to PhP 62.7 billion or 20%.

3.3 Water and land resources utilization for irrigation

3.3.1 Land resources

The land cover statistics are periodically updated by the National Mapping and Resources Information Authority (NAMRIA) based on an interpretation of satellite images. The land cover classification is based on the FAO 2009 International Standards, which has 14 categories: closed forest, open forest, mangrove, annual crop, perennial crop, wooded grassland, shrubs, grassland, built-up area, fallow, fishpond, barrren land, marshland and inland water.

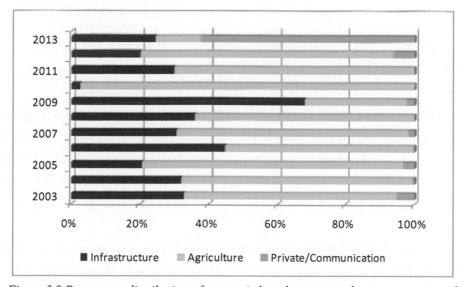

Figure 3.9.Percentage distribution of economic loss due to natural extreme events and disasters, by economic activities (data source: PSA, 2014)

As of 2010 crop land was the largest land cover with about 12.4 million ha or 42% of the total land area of the country planted to annual and perennial crops (Figure 3.10). About 6.8 million ha (23%) was classified as forest and 7.2 million ha (24%) was wooded land, which includes fallow, wooded grassland and shrubland. Built-up areas accounted for 2%.

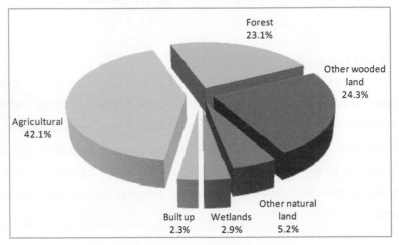

Figure 3.10. Land cover of the Philippines in 2010 (data source: PSA, 2014)

3.3.2   Water resources

The Philippines is endowed with substantial fresh water resources. It has 421 principal river basins, 18 of which are considered major river basins and have drainage areas of at least 1,400 km^2 (Figure 3.11). The major river basins have a total drainage area of 112,500 ha or about 38% of the total land area of the country. There are 79 lakes, 59 of which are natural lakes. The rivers and lakes have a total area of about 1,830 km^2 while freshwater swamps totalled around 1,000 km^2.

In terms of groundwater availability, areas in the country were categorized as shallow well areas, deep well areas and difficult areas. Shallow well areas are characterized by the presence of aquifers or water-bearing geologic formations within the depth of not more 20 metres and with static water level within 6 metres below ground surface. Deep well areas are underlain by aquifers generally located at a depth of more than 20 metres. Difficult areas pertain to areas with minimal groundwater supply and the probability of encountering unproductive aquifer is very high. The country has approximately 52,000 km^2, 123,000 km^2 and 123,000 km^2 shallow well areas, deep well areas and difficulty areas, respectively (National Water Resources Board (NWRB), 1982) (Figure 3.12). The four major groundwater reservoirs include: Cagayan (10,000 km^2); Central Luzon (9,000 km^2); Agusan (8,500 km^2); and Cotabato (6,000 km^2).

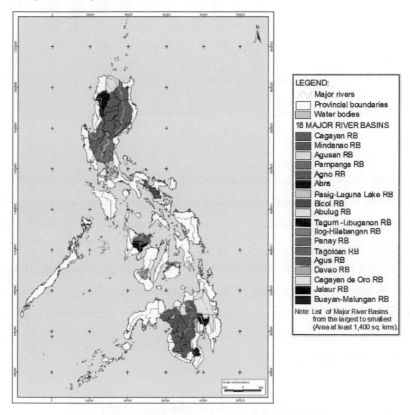

Figure 3.11. Major river basins of the Philippines
(adopted from National Water Resources Board (NWRB)
http://www.nwrb.gov.ph/index.php/products-and-services/water-resources-region-map)

There are 861 impounding dam and reservoir sites for surface water storage. The total available freshwater resources of the country are about 146,900 million cubic metres per year based on 80% probability for surface water.

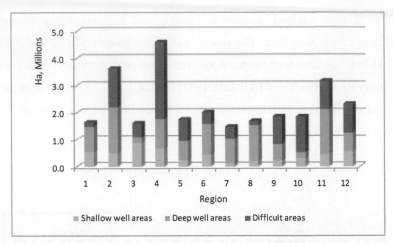

Figure 3.12. Areal extent of groundwater formation in the different regions of the Philippines (NWRC, 1982)

There are no data on actual water abstraction or use at the regional or country level as these are seldom measured. Water allocation or permitted abstraction rates on legally granted water rights is used as an indicator of water use. It is presumably based on demand and is approved by the NWRB.

As of 2013 the NWRB allocated about 199 MCM (million cubic metres) of water or 6,310 m^3s^{-1}. Surface water and groundwater accounted for 98% and 2% of the total volume of allocated water, respectively. Power generation and irrigation sectors had the highest water allocation (Figure 3.13). Except livestock, all water use sectors relied heavily on surface water for their allocation (Figure 3.14).

3.4   Description of irrigation systems of the case study

Balanac RIS and Sta. Maria RIS are small-scale national irrigation systems located in the province of Laguna, respectively about 92 km and 37 km Southeast of Manila - the country's capital (Figure 3.15). Their downstream service areas are in close proximity to Laguna Lake, the country's largest lake. Both systems are under the jurisdiction of the Rizal-Laguna Irrigation Management Office of the NIA Region 4A.

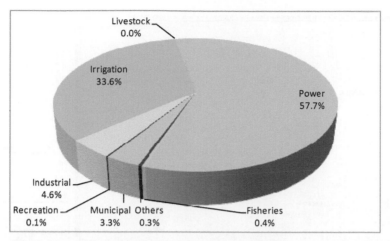

Figure 3.13. Water allocation by water use categories (data source: NWRB, 2014)

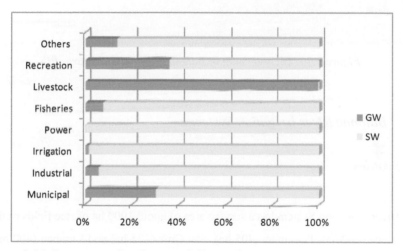

Figure 3.14. Percentage distribution of allocated water, by source and by water use

categories (data source: PSA, 2013)

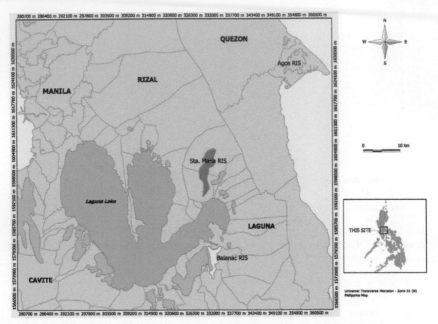

Figure 3.15. Location map of BalanacRIS and Sta. Maria RIS

3.4.1   Balanac River Irrigation System

System profiles

Balanac River Irrigation System has a service area of about 1000 ha of rice fields of the towns of: Magdalena (100 ha), Pagsanjan (405 ha), Sta. Cruz (135 ha) and Lumban (400 ha). It is a gravity type canal irrigation scheme diverting water from the Balanac River. It has a design diversion capacity of 3.9 m^3s^{-1}. The diversion dam is located in Magdalena at North latitude 14° 12' 0.2" and East longitude 121° 26 27.1" and has a drainage area upstream of approximately 116 km^2 (Figure 3.16). The canal network is bifurcating with one main canal, two laterals and two sub-laterals (Figure 3.17). The main canal is 14.9 km long, 87% of which is lined with concrete. About 12.5 km or 82% of the total length of the laterals and sub-laterals is concrete-lined. Two creeks - Salasad and Biñan - serve as the main drainage canal and discharges into the Laguna Lake.

Figure 3.16. Headwork of Balanac RIS

The construction of the scheme started in 1965 with a capital outlay of PhP 0.9 million. It was first operated in 1967. Prior to the operation of Balanac RIS, portions of its present service area used to be irrigated by a few small farmers-built village irrigation schemes that diverted water from Salasad Creek and Biñan Creek. Balanac RIS underwent major rehabilitation in 1984-1989 through the Second Laguna Bay Improvement Project (LBIP) with financial assistance from the Asian Development Bank. The project, which cost about US$ 37.1 Million, included the construction of the run-off-the-river ogee-type diversion dam.

Climate and hydrology

The Balanac River Basin is situated within the Pasig-Laguna de Bay River Basin, one of 18 major river basins of the country. It falls under Type III climate characterized as having no very pronounced maximum rain period with dry season lasting only from one to three months within the months of December to May. The climatological data (Table 3.2) observed at the National Agro-meteorological Station (NAS) of PAGASA in UP Los Baños are adopted for the scheme.

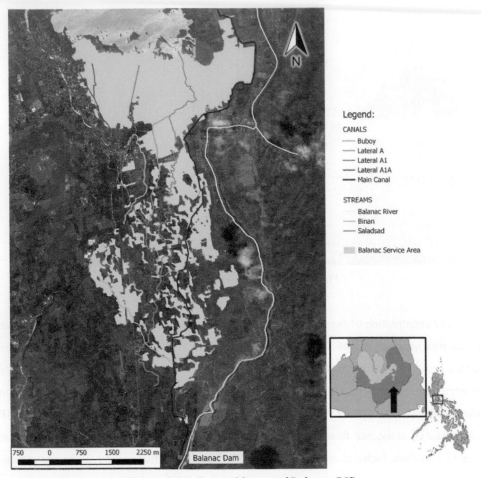

Figure 3.17. General layout of Balanac RIS

For the period 1958-1970, the annual peak flows ranged from 65-531 m^3s^{-1}. The extreme flow values observed were 531 m^3s^{-1} and 0.89 m^3s^{-1}, which were observed in 14 October, 1970 and 27 August 1968, respectively. The computed mean annual peak flow was about 194 m^3s^{-1} (Table 3.3). Mean monthly discharges of less than 9.0 m^3s^{-1} were observed from January to August while greater than 10 m^3s^{-1} were observed for the remaining months (Table 3.4). There have been no flow data for the River since the Department of Public Works and Highway (DPWH) flow gauging activities were discontinued in 1980.

Table 3.2. Normal, average monthly values of pertinent climatologic data, 1985-2015

	Rainfall, mm	Temperature, °C		RH, %	Evaporation, mm
		Min	Max		
Jan	55	21.7	29.6	83.6	101
Feb	35	21.5	30.6	81.8	118
Mar	47	22.3	32.1	79.2	158
Apr	44	23.4	34.1	76.6	180
May	137	24.2	34.4	77.8	164
Jun	223	24.1	33.4	81.6	127
Jul	335	23.7	32.4	83.9	111
Aug	259	23.8	32.2	83.4	115
Sep	253	23.7	32.2	83.7	105
Oct	285	23.5	31.6	83.8	102
Nov	241	23.4	30.9	83.8	92
Dec	202	22.5	29.5	84.5	87

Data source: NAS-UPLB

Table 3.3. Annual discharge Q statistics of Balanac River in m^3s^{-1}, 1958-1970

	Peak Q	Max. daily Q	Mean Q	Min. Daily Q
Mean flow	194	83.9	8.03	2.36
Std. Dev.	136	46.9	3.06	1.14
Skewness	1.3	1.1	-0.8	0.864
Kurtosis	5.5	4.2	3.95	3.47
No. of years	13	9	9	12

Data source: National Water Resources Council (NWRC), 1980

Water management

Balanac RIS is an ungauged system. The flow into the system is controlled by adjusting the main intake structure and the sluice gates at the headwork. The main intake structure is a spindle-type with vertical gates, which are fully closed when irrigation water is not needed, especially during the periods of continuous heavy rainfall or occurrence of typhoons. During periods of low river flows, the sluice gates are lowered to check water and facilitate diversion as the dam is silted with rocks and gravel up to its crest.

Table 3.4. Mean monthly discharges of Balanac River in m^3s^{-1}, 1958-1970

	Mean	Std. Dev	Skewness	Kurtosis	No. of years
Jan	8.76	3.43	0.22	3.62	8
Feb	7.27	3.22	0.53	3.05	8
Mar	6.60	3.48	0.75	3.34	9
Apr	6.01	2.63	0.51	3.40	9
May	5.56	2.76	0.81	4.04	10
Jun	6.08	3.27	2.02	8.30	9
Jul	7.43	3.71	0.98	3.34	10
Aug	8.25	4.14	0.85	4.37	10
Sep	10.6	5.4	1.2	5. 6	11
Oct	12.0	6.4	1.6	6.4	11
Nov	14.0	6.7	1.3	5.8	10
Dec	12.2	3.9	-0.4	3.5	11

Data source: National Water Resources Council (NWRC), 1980

The service area is divided into 34 turn out service areas or management units. The water division at main bifurcating points is through duckbill and proportional weirs. There are direct offtakes, mostly ungated culverts and weir-type, along the main canal and major conveyance canals.

Balanac RIS was under a dual system management by the NIA and the Balanac RIS irrigators association (BRISIA) during the 1990s. Full system operation and management was transferred to the water users association with signing of an irrigation management transfer (IMT) contract between the NIA and BRISIA in 2002. At present, the system is under IMT stage 3 whereby the NIA manages the dam while the water users association is responsible for the operation and maintenance of canals and flow control structures of the canal networks. In general, the NIA is responsible for any needed major rehabilitation and improvement works that cost PhP 50,000 or more, while BRISIA is under the obligation to collect irrigation service fee. The total fee collected is shared in a 70:30 ratio in favour of the NIA.

There are two cropping seasons: the wet and the dry seasons, which start in May and November, respectively. In general the water delivery is continuous. Rotation of water delivery by major bifurcating canal is only practiced during dry seasons that coincide with an El Niño occurrence. System officials of NIA IMO prepare the O&M plan, including cropping calendar and irrigation delivery schedules. These official operation plans are seldom adhered to. In practice, laissez-faire cropping calendar and irrigation diversion are adopted.

Socio-economy in the service area

Based on municipal income, Sta. Cruz, Pagsanjan, Lumban and Magdalena are classified as first-, third-and fourth-class municipalities, respectively (Table 3.5). Agriculture is the main source of livelihood income for many residents. There are 11 other canal irrigation schemes servicing the cropped areas in these municipalities.

Table 3.5. Socio-economic information for municipalities covered by Balanac RIS

Town	Land area, km^2	Land use		Income		Population		
		Agri.	Built-up	Class	PhP x10^6	Total	Urban	Rural
Lumban	96.9	64%	18%	4th	$\geq$ 25; 35	29,470	46%	54%
Magdalena	37.1	94%	2%	4th	$\geq$ 25; 35	22,976	19%	81%
Pagsanjan	26.4	83%	9%	3rd	$\geq$ 35; 45	39,313	100%	0%
Sta. Cruz	38.6	79%	18%	1st	$\geq$ 55	110,943	100%	0%

Data sources: PSA, 2010; Lumban Municipal Planning Office

There are about 1,300 farmers or water users of Balanac RIS. Almost 99% of them are tenants. The farm size ranges from 0.01-21.4 ha with a typical farm size of 0.5 ha (Table 3.6). During the 2012-2015 period, rice yields ranged from 3.8-6.0 ton per ha with an average of 4.3 and 5.5 ton per ha for the wet and dry seasons, respectively. The average production cost during this period was around PhP 35,000 per ha.

3.4.2   Sta. Maria River Irrigation System

Sta. Maria River Irrigation System is irrigating about 974 of rice fields in the towns of Sta. Maria (861 ha) and Mabitac (113 ha). It is a gravity type, canal irrigation scheme diverting water from the Sta. Maria River. It has a design diversion capacity of 2.1 m^3s^{-1}. The two diversion dams are located in Sta. Maria (Figure 3.18): Bagumbayan Dam is at north latitude 14° 30' 34.7" and east longitude 121° 26' 26.4"and has a drainage area upstream of approximately 26.1 km^2; Mata Dam is at north latitude 14° 29' 39" and east longitude 121° 24' 53.9" and has a drainage area upstream of approximately 20 km^2. The canal network is hierarchical with two main canals and seven laterals (Figure 3.19). The total length of the main canals is 13.89 km, 69% of which is lined with concrete. About 14.3 km or 87% or of the total length of the laterals is concrete-lined. The system has nine water reuse facilities that tap drainage water from natural waterways crisscrossing the service areas. All these waterways drain into the Laguna Lake.

Table 3.6. Distribution of farm sizes

Lot size, ha	Number
0-1	1052
>1-2	230
>2-3	41
>3-4	14
>4	10

Data source: Balanac RIS master list

The scheme started operation in 1961. Prior to the operation of Sta. Maria RIS, portions of its present service area used to be irrigated by a few small, farmers-built village irrigation schemes that diverted water from Sta Maria River and numerous small creeks. Sta. Maria RIS underwent its first major rehabilitation in 1976 through the Laguna de Bay Development Project (LBDP) with financial assistance from the Asian Development Bank.

(a)

(b)

Figure 3.18. Headwork of Sta. Maria RIS: Bagumbayan Dam (a) and Mata Dam (b)

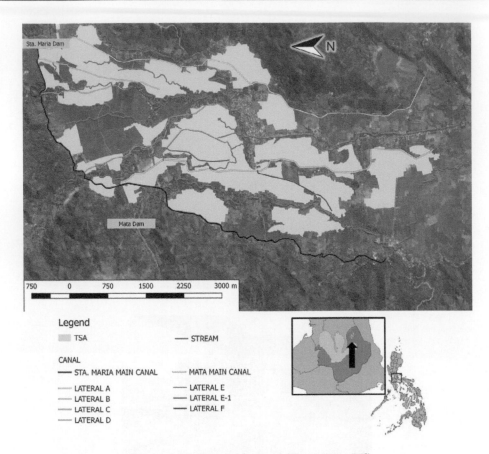

Figure 3.19. General layout of Sta. Maria RIS

Climate and hydrology

The San Antonio River Basin, which includes Sta. Maria River is situated within the Pasig-Laguna de Bay River Basin. Like the Balanac River Basin it is grouped under Type III climate. The nearest PAGASA weather station situated in the town of Pakil, which is at about 11 km distance, was closed down by PAGASA in 2009. Rainfall data (Table 3.7) observed for this station can complement weather data observed at the NAS in UPLB station, which are now used for purposes of irrigation requirement estimation for Sta. Maria RIS. The San Antonio River Basin and the Sta. Maria River are ungauged.

Table 3.7. Average monthly and annual rainfalls for 2000-2008

	Pakil station[1]
Jan	299
Feb	225
Mar	229
Apr	169
May	296
Jun	338
Jul	293
Aug	323
Sep	289
Oct	553
Nov	813
Dec	855
Annual	4580

Data source: PAGASA, 2015

Water management

Sta. Maria RIS is a gated system. Water diverted into the scheme is monitored by the water users association of Sta. Maria RIS at the flumes immediate downstream of the main intakes. The flow into the scheme is controlled by adjusting the main intake structures. The main intake structures of are spindle-type vertical gates, which are fully closed when irrigation water is not needed, especially during periods of continuous, heavy rainfall or expected floods. The sluice gate of Sta. Maria Dam remains fully closed and not-operational while the slab-type sluice gate of Mata Dam is seldom adjusted and is closed most of the time.

The service area is divided into 28 turn out service areas or management units. The flow control structures at main distribution points are a combination of spindle-gated pipe culvert offtakes and adjustable vertical gates as cross-regulators. Constant head orifice structures (CHO) are also used as turn outs. There are direct offtakes, mostly ungated culverts, along the main canals and lateral canals.

Sta. Maria RIS was under a dual system management by the NIA and the Sta. Maria RIS waters users association (SANTAMASI) during the 1990s. Full system operation and management was transferred to the water users association with the signing of an irrigation management transfer (IMT) contract between the NIA and SANTAMASI in 2000. Like Balanac RIS, Sta. Maria RIS is under IMT stage 3.

There are two distinct cropping seasons being followed: the wet season during May-October and the dry season from November to April. Segmental cropping and water delivery schedules are practised in Sta. Maria RIS. The cropping schedules for each irrigation zone follows a system-wide water distribution plan agreed upon by the system officials and water users. Rotation of water delivery is by lateral.

Socio-economic of the service area

Sta. Maria and Mabitac towns are classified as fourth and fifth-class municipalities, respectively (Table 3.8). Agriculture is the main source of livelihood for many residents. There are four other canal irrigation schemes servicing the cropped areas in these municipalities.

Table 3.8. Socio-economic data for municipalities covered by Sta. Maria RIS

Town	Land area, km^2	Land use		Income		Population		
		Agri.	Built-up	Class	PhP 1x10^6	Total	Urban	Rural
Mabitac	89.8	88%	12%	5th	$\geqq 15$; 25	18,618	20%	80%
Sta. Maria	108.4	96%	3%	4th	$\geqq 25$; 35	26,839	0%	100%

Data sources: PSA, 2010; Mabitac Municipal Planning Office

There are about 740 farmers or water users of Sta. Maria RIS. About 53 and 30% of them are tenants and lessee, respectively. The rest were landowners. The farm size ranges from 0.05-23.1 ha with a typical farm size of 1.0 ha (Table 3.9). During the 2012-2015 period, the rice yields ranged from 3.5-6.0 ton per ha, averaging 4.1 and 5.3 ton per ha for the wet and dry seasons, respectively. The average production cost during this period was around PhP 36,000 per ha.

Table 3.9. Distribution of farm sizes

Lot size, ha	Number
0-1	768
>1-2	215
>2-3	42
>3-5	15
>5-10	3
>10-20	1
>20	2

Data source: Sta. Maria RIS master list

4. ISSUES AND CONCEPTS ON DESIGN AND PERFORMANCE OF IRRIGATION SYSTEMS

4.1 Historical perspective on the causes of and solutions to poor irrigation performance

Over the years, the poor performance of many gravity irrigation systems has been attributed to shortcomings in either the design and technology, management and institutional setup, policy or a combination of these factors. Historical perspective on irrigation development from colonial times and turns of related events that led to problems besetting irrigation development in most developing countries today has been discussed by Horst (1998) and Ertsen (2009). Colonial irrigation design and management practices and ensuing irrigation technologies, which were developed to be consistent with the physical and socioeconomic settings and in balance with the management capability of their times, have been continuously adopted in most of the former colonies after World War II. Rehabilitation of most irrigation systems has been merely restoring the original hydraulic design. Meanwhile, foreign technologies have been re-introduced in newer large-scale irrigation systems by project consultants trained on colonial approaches ("schools") of irrigation development. However, many of such technologies have been incompatible with circumstances for irrigation development in post-colonial times. In most cases, local irrigation agencies have to work with limited funds, insufficient staff in terms of number and skills and fewer options for strict control of the irrigation systems. Project planning, design and construction have been isolated from system operation and maintenance. Further, management capacity has not been successfully made to balance with the fast expansion of irrigated areas and is made more insufficient with the growing complexity of system operations.

A number of authors (Horst, 1998; Plusquellec, 2002; Ertsen, 2009) summarized the solutions that had been taken to address these problems. Irrigation systems built in a number of countries during the 1960s consisted only of facilities from the water source to secondary canals. The wastage of water below the farm outlets was viewed as the primary reason for the poor performance of gravity irrigation systems. In the 1970s, the common responses included tertiary unit development (increasing the density of canal systems including introduction of modern water application methods and precise land levelling), technical training of farmers, operators and extension workers; creation of water users groups at the level of tertiary canals and refining of flow measuring techniques. However, these conventional engineering solutions failed to push the performance of irrigation to expected levels. In the 1980s, the focus of efforts to improve the performances of irrigation shifted to the management and institutional aspects. To date, the results left much to be desired.

Several authors (Rice, 1997; David, 1997; Horst, 1998; Plusquellec, 2002; David, 2003) argued that over-optimistic assumptions in the planning stage and faulty and unrealistic designs were the main causes of the poor performance of gravity irrigation systems. The results of the 1996 and 1997 studies of the Operations Evaluation Department (OED) of the World Bank on six gravity irrigation systems in Thailand, Myanmar and Vietnam agreed with this opinion. Findings from this study contradicted the dominant paradigm attributing low economic returns of government-operated, gravity-fed irrigation schemes to poor O&M and inadequate farmer organizations. The study concluded that the primary reasons for performance gaps included over-optimism about the crop area to be served and project design faults including the choice of unsuitable technology.

The recognition of poor design as a main reason of the poor performance of irrigation systems has been slow. Plusquellec (2002) summarized the studies, publications and events that acknowledged directly or indirectly the importance of design and technology in the performance of irrigation systems. He noted that although such studies, publications and events were sponsored by international organizations (World Bank, International Water Management Institute (IWMI), International Commission on Irrigation and Drainage (ICID) and FAO, they remained the initiatives of individual experts rather than the result of policy adopted by such organizations. Consequently, improvement of the performance of irrigation projects through revisioning of designs was not given a high priority on the agenda of international fora on water.

Clemmens (2006) deduced that the main reason for the poor performance of large-scale canal irrigation systems is that system operation is not tied up to the productivity of irrigation systems. He argued that chaos, which he defined as anything that causes the process within a system to be variable and difficult to predict, naturally dominates large-scale canal irrigation distribution systems. The dispersive nature of these irrigation systems, even those with reasonable infrastructure and management, causes significant variability in water supplied and water delivery service, especially at the downstream of the delivery system. Shortcomings in design, maintenance and operation make distribution and irrigation water losses worse. He put forward that the management philosophy of canal irrigation systems should adopt a service orientation. His proposed remedy is to improve both the physical and administrative controls at key intermediate points within the distribution network in accordance with such service philosophy of management. The improvement of physical controls relates to modernisation of irrigation infrastructures, that is, installing modern flow control structures needed to isolate lower parts of the canal network from upstream disturbances and chaos. Administrative controls include among others the establishment of water delivery criteria that are agreed upon by the water users and the service providers, a contract stipulating the accountability of the service provider and willingness to pay for the service, and purposeful corrective actions to overcome chaos.

The modernisation process of irrigation systems is now widely recognized as a strategic approach to improve water productivity and increase total production and economic returns of large canal irrigation schemes (Burt and Styles, 1999). In the past decade, the concept of modernisation has evolved from merely introduction of new physical structures and flow control devices to a process of technical, managerial and institutional transformation of irrigation systems with the objective of improving resource utilization and water delivery service to farms (Plusquellec et al., 1994; Plusquellec, 2002; Playan and Mateos, 2006). Ertsen (2009) noted the wide range of interventions or efforts associated with modernisation, from increasing drainage intensity to precise amount and timing of water delivery.

Recently, FAO put emphasis on modernisation as a process of improving the ability of the irrigation system to serve the current demands of farmers as well as to provide flexibility to responds to future needs with the best use of the available resources and technologies (Renault et al., 2007). Its irrigation modernisation programme in at least 30 irrigation systems in Asia highlighted inadequate attention to canal operation as a major reason for dismal performance of canal irrigation systems. The many surveys carried out by FAO disproved the then widely taken belief that the issue of underperformance of irrigation systems is related to socio-economic factors and not to engineering aspects. It showed that substandard canal operation is very often the origin of the vicious cycle of poor irrigation service, poor fee recovery, poor maintenance resulting in the deterioration of the physical structures and water delivery service (Renault *et al.*, 2007). In connection with this, the FAO presented a methodology to assist water engineering professionals, irrigation managers and practitioners in developing a consistent modernisation plan for medium- to large-scale canal irrigation systems. This methodology is called MASSCOTE (Mapping System and Services for Canal Operation Techniques), a stepwise procedure for diagnosing the performance, assessing the different system characteristics and mapping the way forward in order to improve the irrigation service and the cost-effectiveness of canal operation. Its focus is on canal operation, but the scope is modernisation and the goal is to promote service-oriented management of irrigation systems (Renault *et al.*, 2007).

The merit of service orientation of management, performance assessment and design approach for modernisation of irrigation systems has been pointed by a number of authors. Good irrigation service provides the environment for productive and sustainable agriculture (Molden *et al.*, 2007). Service orientation in the management as a key element of the strategy that is needed to improve the performance of canal irrigation systems is shared by other authors (e.g.: Clemmens, 2006; Malano and Hofwegen, 2006; Renault *et al.*, 2007). Ertsen (2009) argues that farmer interventions, which often reflect a nature of water delivery service demanded by farmers, should be taken as standard in irrigation design. A service-oriented approach implies that design engineers should select the physical elements of the system within a clearly defined operational strategy that will ensure delivery of the water service demanded. Service concept has been the basis of modern design (Plusquellec, 2002).

Meanwhile widespread emergence of private tubewell irrigation systems, either as a supplemental irrigation or a sole source of irrigation water in several Asian countries, has occurred since the early 1980s (IPTRID, 2003; David, 2003; Turral *et al.*, 2010). The rapid growth of groundwater irrigation in these countries is viewed as a farmers' response to low level of flexibility and reliability of canal irrigation systems (Plusquellec, 2002). The popularity of private tubewells among farmers is mainly attributed to a higher reliability of groundwater supply, higher level of farmer's control over tubewells and greater flexibility in planning their cropping pattern (Plusquellec, 2002; David, 2003; IPTRID, 2003; Hussain, 2007b). However, consequences of overexploitation of groundwater resources, such as alarming overdraft in groundwater level and decline in water quality, have been increasingly observed in many areas.

4.2 Design issues

The common design problems resulting in poor performance of irrigation systems were pointed out by a number of researchers. Meijer (1992) cited three common pitfalls in irrigation design. These are:

- the use of too optimistic friction factors when applying the Manning formula in small unlined canals;
- the unbalanced design procedure that largely "follows the direction of the water flow";
- the adherence to high water use efficiency.

The Manning's equation is expressed as:

$$v = k_M R^{2/3} S^{1/2} \tag{4.1}$$

where v is the flow velocity in m s^{-1}; k_M is the flow factor related to the canal roughness and is also known as the friction factor $1/n$; R is the hydraulic radius, or the quotient of water cross-sectional area and wetted perimeter in m; and S is the energy gradient or the bed slope for canals with normal flow. Meijer pointed out that for uniform flow in rough canals, the canal roughness or friction factor is not a constant but increases with the depth of flow. He opined that the variations in the canal roughness are probably due to the simplification of the formula by adopting 2/3 as an average exponent of the hydraulic radius R when in the derivation of the formula the exponent was found to vary from 0.65 to 0.84. The canal roughness is also determined by the canal shape and bed slope. The influence of depth of flow, canal shape and bed slope is more pronounced in rough shallow sections or at small depths of flow. Meijer concluded that this simplified Manning formula leads to too small dimensions in small earthen canals to convey the design flow. He proposed the use of a modified Manning formula of the form:

$$v = k_M y^{1/3} R^{2/3} S^{1/2}$$ (4.2)

where y is the flow depth. He suggested a k_M value that equals 36 ($n = 0.028$) for well maintained irrigation canals with weeds that are cut to 0.05 m and depth of flow not more than 3 m. In small open irrigation canals carrying less than about 0.2 $m^3 s^{-1}$, a k_M value of 24 ($n = 0.042$) is recommended.

The second technical shortcoming referred to by Meijer (1992) is a sequence of the design process where the irrigable area and the alignment of the main canal are considered first, followed by the secondary canals, roads, and drains, and then by the tertiary units. Usually the tertiary units or farm blocks were designed last, often long after the main and the secondary canals and roads had been constructed. Such design succession dates back from the 1940s when irrigation design mainly concentrated on ensuring water supply to groups of farmers already drawing water from less reliable sources. Meijer argued that problems with such a design sequence arise after the implementation of the irrigation scheme if during the design the details of the tertiary unit including the layout are not known, or not considered.

For the third pitfall, Meijer argued that too much adherence to high water use efficiency per se in the design of irrigation systems can result in water distribution schedules that tend to be much too complicated and far too rigid for everyday practice. He contended that apart from crop water requirements, additional water is needed to facilitate a fair and simple water distribution. These so-called additional operation requirements (AOR), management losses or intentional losses are necessary for simplifying the water distribution and system operation.

Meanwhile, Pluesquellec *et al.* (1994) pointed out that the design and layout of an irrigation scheme often fail to consider some basic laws of hydraulics, such as lag time, unsteady nature of water flow, and fluctuations of the water level. All too often the designer assumes that the canal will operate well with unsteady flow, but in reality the design prohibits effective operation because it lacks a control strategy, sufficient communications, suitable gate spacing, or other design errors. One of the most common design errors is the use of manually operated undershot gates in cross-regulators where almost invariably the function of the structure is to simply maintain a relatively steady upstream water level while passing flow changes on to the next reach.

Pluesquellec *et al.* (1994) traced other design errors to a poor understanding of the interactions of a particular hydraulic structure with other structures in its vicinity. Three common problems relate to location and interaction of structures. First, many conventional canal designs have too few check structures to ensure adequate water-level control at outlets. Thus, the flow rate through the outlets varies with time even though it is supposed to remain constant. Second, the difference in head across structures is frequently insufficient. Larger changes in elevation will dramatically stabilize the flow rate at standard underflow turnouts. However, drops across farm turnouts recommended in some design manuals are insufficient. Third, the designs of lower-level (tertiary, distributary and watercourse) distribution systems are frequently not based upon sufficient topographical data, with the result that the supply system may be at a lower elevation than the destination of the water. They opined that some structures, which are still found in new irrigation systems, have either undesirable hydraulic effects or are premised on unrealistic hydraulic assumptions. These include Romijn gates, vertical sliding gates as check structures, radial gates as check structures without side weirs, and constant-head orifice offtakes. The weir action of Romijn gates creates large fluctuations of the flow rate in the offtake with minor changes in the water levels of the supply canal. Sluice-gated control along the parent canal combined with Romijn-gated offtakes and outlets is among the unfavourable combinations from hydraulic point of view, since it is extremely unstable and requires high construction standards, accurate calibration, and disciplined operation if correct discharges are to be maintained. Small deviations have a proportionally great effect. On the other hand, vertical sliding gates are underflow gates, therefore the upstream water level tends to fluctuate widely with small adjustments. Radial gates as check structures without side weirs should be avoided in manually operated systems because they are basically underflow weirs with the same problematic features as sliding gates. The basic hydraulic assumption for constant-head orifice offtakes is a stable water level in the supply canal or relatively insignificant fluctuation between adjustments. But since this assumption is often not the case, adjustments on the offtakes must be made. Such necessary adjustments make operation relatively complicated.

Further, Plusquellec *et al.* (1994) argued that many failures and problems are caused by a design approach that pays insufficient attention to operational procedures. The designs of many canal irrigation systems make management of these systems difficult in actual field conditions. Instructions for system operations are often conflicting and sometimes meaningless. They cited Murray-Rust and Snellen (1991) observation of an operation system of the Maneungteung Irrigation Project in Indonesia that requires bi-weekly assessment of demand for every tertiary block and readjustment of every gate in the system to meet the changed water distribution plan, thus requiring a very intensive data collection program and an efficient and effective information management system. Because such operational procedure is carried out in an environment of unpredictable water availability, it becomes almost impossible to achieve.

Citing the work of other researchers, Plusquellec *et al.* (1994) mentioned the following examples of design problems:

- an actual operation procedure for the Right Main Canal of the Kirindi Oya Irrigation Project in Sri Lanka would cause a water fluctuation of about one metre in the lower canals and take up to four days to reach a steady state;

- gate rating tables, which tell operators how much to open a cross-regulator for specific discharges, are over-emphasized. They are generally inaccurate. For typical cross-regulators in a series in a canal, the use of gate rating is meaningless. With upstream control, the flow rate through the structure cannot be controlled by the operator except for a short duration because the flow in the canal is determined at the canal inlet. Instructions for cross-regulators with upstream control should deal with maintaining a desired water level, not a desired flow rate;

- manually operated gates and control structures rarely work despite all efforts to improve irrigation management and the capacity of staff;

- many extremely simple irrigation systems are so rigid that the operational procedures cannot be changed when required. The design must include sufficient flexibility to adjust the operation procedure.

Plusquellec *et al.* (1994) reasoned that failures of some modernized projects were due to improper choice of control structures, incompatible components, and a design that was not based on realistic operation and maintenance plans. These gave a wrong impression that modern design concepts are not suitable to emerging or least developed countries.

In addition, Plusquellec (2002) cited the hierarchy of canals as one specific design issue that is neglected during the design of irrigation projects. In many existing schemes in South Asia, as much as 40% of the irrigable area in some projects is served through direct outlets from the branch distributary and even from the main canal. He argued that such canal arrangement is hardly amenable for the creation of multi-tiered users associations which would be useful in the effective management of the system, lack of which has also detrimental effects on the system performance.

Horst (1998) believed that the problems in irrigation are historically determined, finding their roots in the technology of the colonial era and the lack of adaptation to the new socio-economic environment of the post-colonial period. During the colonial times, design and operation resided in one ministry making feedback from operation to planning and design possible. Consequently, technology developed in balance with the management capability. In the post-colonial era, however, design has been carried out by foreign consultants while government agencies have been responsible for operation. Horst argued that this separation of design and operation has led to discrepancies between design assumptions and operational realities. Together with shortage of technical personnel and expansion of areas under irrigation, irrigation in emerging and least developed countries has been haunted for decades by a multitude of problems such as low performance, low irrigation efficiencies, conflicts among farmers and between farmers and management and deteriorating physical structures.

Similarly, Horst concluded that often in the design phase, little or no attention was paid to operational aspects. Design assumptions were limited to agronomic, engineering and economic parameters only, without taking into consideration institutional and human aspects. Thus, the outcome of the design was more often incompatible with the socio-institutional environment. Horst specifically identified the design of water division structures at the core of many irrigation problems. Owing to their functions, he argued that the type and characteristics of these structures largely determine the operability and manageability of the system and are the points of interface where conflicts of interest among farmers and between farmers and management often take place. He pointed out that not only the design of many water division structures was based on incorrect hydraulic supposition but was also devoid of social and institutional criteria, such as staff requirement (in terms of number and skills), operability and social acceptance, among others. Such design shortcomings led to hydraulically unstable canals that were too cumbersome to operate, thus, required extra field staff. Horst discussed the common types of water division structures in terms of general hydraulic characteristics and operational implications. He concluded that the types of irrigation systems with manually or mechanically operated water division structures experience most serious problems.

Cases of canal irrigation systems with unsound design philosophy and design criteria were also observed in the Philippines (Comprehensive Irrigation Research and Development Umbrella Program, 2007). The results of a preliminary study of sample canal irrigation systems in the Province of Ilocos Norte showed numerous flawed design parameters (David, 2009). These include lack of head control structures, wrong placements and combinations of flow control and offtake structures, improper alignment of main farm ditch and farm ditches, ungated offtake structures, direct offtaking of farm ditches from main canals, inadequate protection of sluice gates and main intakes from siltation and remarkably high service area to farm ditch and turnout ratios. Potentials for improvements in hydraulic structures were evident. However, the most serious design mistake was the gross underestimation of percolation and seepage during the design stage. There was also an apparent underestimation of flood flows for one large diversion dam and one impounding dam. These dams were designed for 100-year and 25-year floods, respectively, but were both badly damaged by a typhoon whose rainfall totals had an associated chance of occurrence of 25% per year. Unfortunately, such mistakes were most likely repeated in other irrigation systems in the country for one or combinations of the following reasons:

- only one national agency is involved in implementing gravity irrigation projects;
- design criteria are generalized over regions or larger areas;
- absence of feedback mechanism between design and operation engineers.

As cited by Plusquellec *et al.* (1994), similar design mistakes were observed by Murray-Rust and Vermillion (1989) in Indonesian irrigation projects. These errors included the following:

- over- or under-estimation of water availability for irrigation, both during the year and various seasons;
- wrong estimates of field level demand for irrigation due to incorrect design assumptions of field efficiency, cropped acreage, and cropping pattern;

- incorrect conveyance losses used in sizing and locating unadjustable structures rendering such structures useless in actual system operation.

According to Plusquellec (2002), the control of seepage losses is one design issue that is generally neglected during the design of irrigation projects.

4.3   Concepts on irrigation design

Noting the traditional design sequence of irrigation projects wherein the tertiary units or farm blocks are designed last, Meijer (1992) recommended a design sequence wherein the so-called bottom-up and top-down design approaches are employed (Figure 4.1). The sequence of the first two steps is arbitrary and the procedure is repeated several times until a satisfactory final design is achieved. The conclusions arrived at in step 3 are used in steps 1 and 2 of the next round. The top-down design is mainly used when moving from step 2 to step 3 while the upward approach is apparent between steps 1 and 3. Any inconsistencies between the two approaches should be solved in step 3.

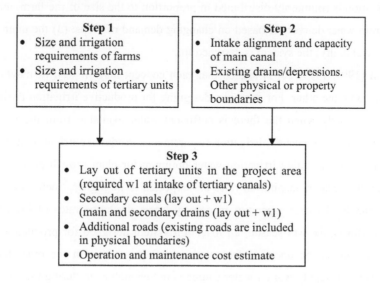

Figure 4.1. Simplified conceptual framework of the research

Ankum (2001) pointed out that large-scale canal irrigation systems under dual management can only be well designed and managed when their overall objectives have been specified accurately, together with the resulting operational objectives of the main system. He refers to the main system as the part of a dual-managed irrigation system that is under the control of an irrigation agency and usually comprises the headwork and all canals and appurtenant structures up to the tertiary offtake. The tertiary offtake is the structure that diverts water from the main system into tertiary units, the part of the irrigation system that is presumably under the control of a farmers' group. He emphasized that the design team (e.g.: engineers, agronomist, economist) and users of the irrigation system (e.g.: irrigation agency, farmers) have to jointly agree on the stipulated objectives. Otherwise, arguments on whether the irrigation system is performing according to expectation will remain.

He also made distinctions between the irrigation system's overall objective and the main system's operational objectives. Examples of overall system objectives include irrigation concepts such as: (1) protective irrigation (also called "supply-based" irrigation) wherein the limited water supply is distributed over the entire service area for drought or famine relief, and productive irrigation (also known as "demand-based" irrigation) that focuses on the satisfying the irrigation water of the crop for optimum production; (2) equitable supply wherein irrigation is commonly distributed in proportion to the size of the farm, and flexible supply wherein water delivery is based on changing demand of crops; (3) the main irrigation season (e.g.: wet or dry) the system is designed for.

Ankum (1997; 1999; 2001) argued that each concept may fit well or maybe the only logical match with the other. For example, designing for productive irrigation during the dry season is logical only when the there is sufficient water available from the water source during the dry season to meet the irrigation requirements. Productive irrigation during the dry season would mean providing irrigation water optimum for plant growth during a cropping season when crop evapotranspiration rates are expected to be higher. Such objective would translate to canals of larger capacities to convey larger discharge rates of irrigation water (Figure 4.2). During the wet season, an irrigation system designed for productive irrigation during dry season can be operated for supplemental irrigation. On the other hand, when rainfall or irrigation supply from a water source is not enough even during the wet season or monsoon months, designing for productive irrigation would imply a focus on providing supplementary irrigation during the wet season. In this case, medium capacity canals are constructed.

Meanwhile, the design context of protective irrigation is that the supply available at the water source is inadequate to meet the irrigation requirements of the cropped area. The focus of protective irrigation is on distributing the available water to as many farmers as possible. With this context, the irrigation system is designed for wet season irrigation as designing for dry season is not a logical option since the latter case would results to higher irrigation water requirements that would be impossible to meet due to the given water supply constraints in the area. Canals for protective irrigation are usually of small capacity.

Equitable supply is applied in protective irrigation systems where all farmers receive their share of water (Figure 4.3). It may also be applied in productive irrigation for a uniform cropping pattern. Flexible supply fits well in productive irrigation systems with diverse cropping patterns. He also opined that reshaping an existing protective irrigation system serving rice areas into productive system would not be feasible without drastically increasing canal capacities and water supply availability.

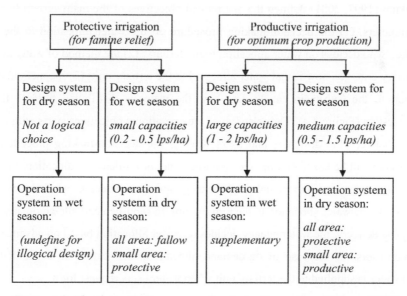

Figure 4.2. The determining season of an irrigation system (Ankum, 2001)

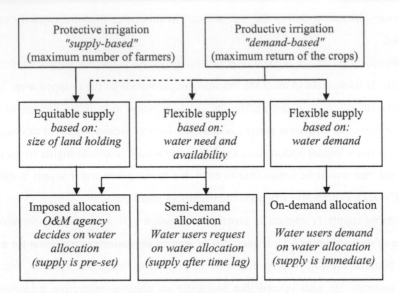

Figure 4.3. Equitable and flexible supply and decision-making procedure for an irrigation system (Ankum, 2001)

Ankum (1997; 2001) defined the operational objectives of the main system in terms of three parameters: (1) the decision-making procedure on the water allocation to the tertiary offtake, i.e., who decides on water allocation to the tertiary; (2) the method of water allocation to the tertiary unit, i.e., how is that water allocated to the tertiary unit; (3) the method of water distribution to the (sub)main canals, i.e., how the water is distributed through the main irrigation system (Figure 4.4). The selection of the operational objectives starts with a general agreement between the designers and users whether the decision-making procedure at the tertiary offtake will be based on imposed, semi-demand, or on-demand allocation. In imposed allocation, also known as dictated delivery, the irrigation agency decides while the water users have no say on the water allocation to the tertiary unit. In semi-demand allocation, also known as on-request delivery, the water agency decides on water allocation based on advance request from water users. In the case of on-demand allocation, also called on-demand delivery, irrigation water is supplied to the tertiary unit when water users request for it.

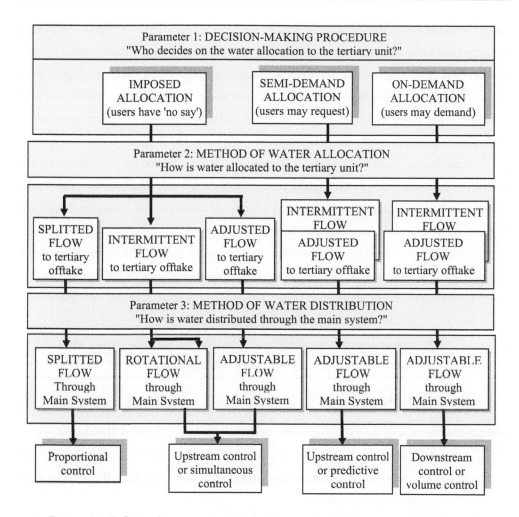

Figure 4.4. Relation between parameters of operational objectives and the logic choices for flow control methods (Ankum, 2001)

Further, he pointed out that the decisions on protective or productive irrigation and on irrigation cycle, that is, the timing of gate settings at the tertiary offtake determine already the decision-making procedure to the tertiary offtake. Protective irrigation wherein no water request is entertained fits well with imposed allocation. In contrast imposed, semi-demand and on-demand allocation can be applied with productive irrigation. The irrigation system is under imposed allocation when gate adjustments at the tertiary unit are not possible. If gate adjustments are possible at any time, it is under on-demand allocation. When gate adjustments are only possible at fixed times it is either under semi-demand or imposed allocation. Further, the decision-making procedure on water allocation to tertiary units (i.e.: imposed, semi-demand, on-demand) directly connotes the timing of water supply delivery (i.e.: pre-set, after time lag, immediate). The rates of supply delivery can be surmised to be according to the definition of productive and protective irrigation.

According to Ankum (2001), the consensus on the decision-making procedure on water allocation determines already the options for the methods of water allocation to tertiary units and the water distribution through the main system. He identified the following three basic logic methods irrigation water can be allocated to a tertiary unit:

- splitted flow or also called proportional flow, when the discharge from the headwork is diverted into a fixed ratio throughout the main systems and towards the tertiary units;
- intermittent flow or also called on/off flow, when water is allocated to tertiary units either at a maximum flow (peak discharge) or at zero-flow;
- adjustable flow or also called varied flow, when the discharge to tertiary units is both continuous and can be intentionally adjusted to the target rates.

His list of options for logical combination of the three fundamental parameters into an operational objective of the main system (connected by arrows in Figure 4.4) includes:

- imposed allocation to the tertiary offtake by splitted flow allows splitted flow through the main system. An intermittent, rotational or an adjustable flow through the main system are also possible options;
- imposed allocation to the tertiary offtake by an adjustable flow allows a rotational flow or an adjustable flow through the main system;
- imposed allocation to the tertiary offtake by an adjustable flow prescribes an adjustable flow through the main system;
- semi-demand allocation to the tertiary offtake by an intermittent flow and by an adjustable flow prescribes an adjustable flow through the main system;
- on-demand allocation to the tertiary offtake by an intermittent flow and by an adjustable flow prescribes both an adjustable flow through the main system.

Further, Ankum (1997; 2001) pointed out that the parameters of the selected operational objectives of the main system limit the options for flow control method and the type of flow control structures (that is, water level and discharge regulators) and their operational instructions. Flow control methods aim at maintaining a target water level or target discharge. The logic choices for the flow control method (Figure 4.4) are as follows: (1) either proportional control or upstream control for imposed allocation; (2) normally upstream control with central management, but predictive control under self-management for semi-demand allocation; (3) either downstream or volume control for on-demand allocation.

He emphasized that the selection process of flow control method requires a multidisciplinary approach, but ultimate selection should be closely related to the operational objectives of the main system. The six basic flow control methods based on the location of the target value, also called setpoint, are described as follows:

- proportional control, without any setpoint in water level. The primary purpose of a structure is to maintain a certain ratio-in-discharges to downstream destinations. Irrigation systems under proportional control divide and distribute the water according to a fixed ratio.

- *upstream control*, with the setpoint in water level at the upstream side of the regulator. Structures under upstream control maintain the upstream water level at the target level, which means that the upstream water level is constant for each discharge. A main irrigation system under upstream control is essentially a discharge-controlled system, that is, discharge regulation and measurement is required at all offtaking canals. It requires a central management where the irrigation agency decides on the water releases to tertiary units and on the cumulated discharges through the whole main system;.

- *downstream control*, with the setpoint in water level downstream of the regulator. It is a control of the water regulator based upon changes in the water level immediately downstream of it. The regulators in the main system maintain a constant water level downstream of the structure, regardless of the discharges. Such regulation of structures means that water supply is given to a canal reach when the water level drops. Thus, the discharge at each regulator is automatically adjusted to the accumulated downstream demand for irrigation water. A downstream-controlled main system has self system management, that is, the system itself converges to the new (required) steady state;

- *volume control*, with the setpoint in water level at a pivot point located in the middle of the downstream reach. Its concept is to have a constant volume, regardless of the discharge in the downward canal. Volume control behaves basically as a downstream-controlled system. It solves the large dynamic storage in traditional downstream control but requires telemetry and regulation by electro-mechanical gates;

- *predictive control*, with the setpoint in water level at the other end of the downstream canal reach. It uses two sensors per regulator: the main sensor is located at the next downward gate to measure water need there; the second sensor is at the gate itself as to prevent overtopping when irrigation flow stops. Predictive control solves the drawback of central management in upstream control and avoids the need for high embankments for the positive wedges in downstream control and volume control, but suffers the same problems with time-lags and operational losses associated with upstream control and requires telemetry and regulation by electro-mechanical gates;

- *simultaneous control*, without setpoint water level, but with discharge regulators in all canal reaches. A system under simultaneous control behaves basically as a piped system without dynamic storage

Plusquellec *et al.* (1994) argued that there is a pressing need to adopt a modern or new approach to the design and engineering of irrigation projects to improve the performance of canal irrigation systems. Modern irrigation design is defined as a result of a thought process that selects the configuration and the physical components in the light of a well-defined and realistic operational plan, which is based on the service concept. It is not defined by specific hardware components or control logic, but by the use of advanced concepts of hydraulic engineering, irrigation engineering, agronomy and social science to arrive at the most simple and workable solution.

Further, they suggested that engineers must follow a certain evolution in their thoughts and in their perception of the design task in order to produce a good final product. The steps in this evolution and the associated requirements are given in Table 4.1. The classical approach tends to concentrate on physical components without considering their interactions. The focus is on the mechanical and structural integrity. The first step is associated with most of the design mistakes.

Table 4.1. Evolution of irrigation project designers (Plusquellec *et al.*, 1994)

Step	Design task	Requirements
First	Classical	Degree in engineering; access to design handbooks
Second	Functionality of design	Some understanding of unsteady flow; ability to comprehend the function and purpose of various control devices including semi closed pipelines; some field observation to verify the actual function of each component
Third	Robustness, verification of equity	Exposure to a wide variety of structures; good understanding of flow in open canals; a deep appreciation for simplicity; understanding of the difference and importance of water-level control in some cases and flow-rate control in others
Fourth	Ease of operation	Good understanding of and experience with operation of multiple structures with unsteady flow; extensive exposure to a variety of operational procedures and structures, however, a detailed knowledge of unsteady flow equations is not necessary; sympathy for harried operators and farmers
Fifth	Consideration of water balance and quality issues in the hydrologic basin	Exposure to broad issues in water management; understanding of true definitions of irrigation efficiency at various levels within a hydrologic basin
Final	Recognition of typical design errors	Experience and critical evaluation of one's own previous design; experience with operation of irrigation projects, including impartial analysis of anticipated operation in contrast to actual operation and constraints

David (2007) maintained that appropriate design criteria are prerequisites to good design, and hence, engineering performance. Design criteria are the parameters on which the specifications for structures and their construction, operation and maintenance systems are based. They are determined usually from hydrologic, geologic, agronomic, climatologic and socio-economic characteristics of the project area. They are influenced by design philosophy, which takes into consideration, among others, national and regional policy environment, sustainability, cost-effectiveness, environmental impact, social equity, management and cultural factors. Examples of design philosophy include manual construction as against mechanized construction to generate local employment, use of locally available construction materials, flexibility for crop diversification, empowerment or greater degree of control of farmers over management of irrigation water, continuous or rotational irrigation, and simultaneous or relay cropping. The first crucial point in the design of an irrigation system is a clear understanding of, and formulation of a design philosophy and determination of design criteria. Table 4.2 lists the common design criteria and the relevant parameters for some canal irrigation systems.

For Horst (1998), the water delivery schedule or the way in which water is delivered at the tertiary unit is the core of the design of irrigation systems. It is derived from a number of choices and assumptions both at field and system levels. Design assumptions deal partly with policy planning (extensive or intensive irrigation, poverty alleviation or production-driven, type of water allocation, etc.) and partly with agronomic requirements (cropping patterns, water requirements, irrigation methods, etc.). Figure 4.5 shows the conventional derivation of the water delivery schedule. The upper section of the chart represents the design decisions, which are concerned mostly with the agronomic and agro-hydrological aspects of irrigation at the farm level, while the lower section deals with design choices at system level. The latter are primarily related to technological and operational aspects. Matching the two sections will result in the water delivery schedule. Once the water delivery schedule is determined the system technology follows as a derivative from this schedule. In its turn, the type of system technology determines the mode of operation of the system.

Table 4.2. Common irrigation design criteria and the parameters influencing them (David, 2007)

Type of irrigation system	Water diversion or intake at the source		Conveyance and distribution of irrigation water		Water utilization at farm level	
	Design criteria	Relevant parameters	Design criteria	Relevant parameters	Design criteria	Relevant parameters
Run-of-the-river diversion dam	dependable water supply	streamflow records	Capacity	canal dimensions	irrigation water requirement	effective rainfall, crop water requirement
	Flood	streamflow records (peak flows)		irrigation water requirement, application efficiency	application efficiency	
	diversion water requirement	service area, farm water requirement, conveyance efficiency	conveyance efficiency	seepage and percolation		seepage and percolation in tertiary canals
	water quality	type of crop, water quality requirement by crops	canal bed width to water depth (b/d) ratio	canal dimensions	crop water requirement	type of crop, soil type, percolation and seepage within farm, evapotranspiration
	command head	water surface elevation, farm elevation, head losses	maximum radius of curvature side slope	canal/bank stability soil/canal lining properties		

Modernisation strategy for the national irrigation systems in the Philippines

Table 4.2. *Continued*

Type of irrigation system	Water diversion or intake at the source		Conveyance and distribution of irrigation water		Water utilization at farm level	
	Design criteria	Relevant parameters	Design criteria	Relevant parameters	Design criteria	Relevant parameters
	sediment load	siltation rate	maximum permissible velocity	canal lining		
			measuring and regulating structures	intended purpose, location, ease of operation		
	stability of barrier/ dam	barrier/dam dimensions, design flood, spillway capacity (for diversion dam)	maximum canal slope	soil/canal lining properties		
Small water impoundment	dependable water supply	inflow-outflow hydrograph, reservoir storage (active and flood), evaporation, seepage and percolation	Capacity	canal dimensions	irrigation water requirement	effective rainfall, crop water requirement

Issues and concepts on design and performance of irrigation systems

Table 4.2. Continued

Type of irrigation system	Water diversion or intake at the source		Conveyance and distribution of irrigation water		Water utilization at farm level	
	Design criteria	Relevant parameters	Design criteria	Relevant parameters	Design criteria	Relevant parameters
Flood		rainfall records, runoff coefficient (catchment area)	farm water requirement	irrigation water requirement	application efficiency	seepage and percolation in tertiary canals
	command area	water surface elevation, farm elevation, head losses	conveyance efficiency	seepage and percolation in canal	crop water requirement	type of crop, soil type, percolation and seepage within farm,
	sediment inflow	siltation rate	measuring and regulating structures	intended purpose, location, ease of operation		evapotranspiration
	water quality	type of crop, water quality requirements of crops				
	pond capacity	pond area, embankment height				

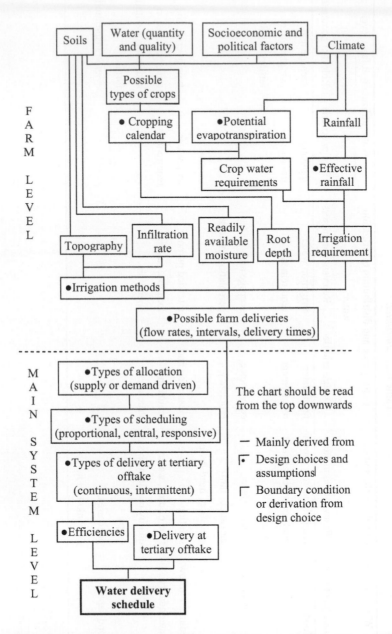

Figure 4.5. Conventional derivation of the water delivery schedule (Horst, 1998)

On top of agronomic, engineering and economic design criteria, Horst (1998) proposed the explicit inclusion of institutional and human aspects such as operability, operational procedures required, staff requirements, transparency, corruptibility, social acceptance and farmer management to the list of criteria or consideration in the design. He proposed simplification of operation and technology by simplifying the water delivery. He examined the applicability of simplified technologies such as proportional outlets, proportional division weir-type structures, on-off gates and stepwise distributors in six commonly recurring cropping patterns. He concluded that for many different cropping patterns a simplified technology (either proportional division or on-off structures) is applicable provided additional operational requirements are taken into account. While admitting that additional operational requirements can be considered as a loss of water, hence, contributes to low irrigation efficiencies, Horst (1998) argued that with this intentional management loss, the overall water use efficiencies are expected to increase due to a simpler technology and a simpler operation. As Meijer (1992) has put it, one needs water to save water.

Renault (2001) advocated a concept of re-engineering irrigation management and system operation, which combines the notion of industrial production (transformation of inputs into outputs) and the notion of service. He viewed re-engineering as a revolution in the thinking and design concerning irrigation management and operation. Thus, it should be the first strategic stage of any intervention and must be a prerequisite to any physical (modernisation of infrastructure) or institutional (reforms) interventions. With the re-engineering concept, he considered a re-assessment of some of the commonly held assumptions underpinning irrigation design as a likely first step in identifying new ways of managing irrigation systems. He proposed to discard the assumption on homogeneity in cropping patterns and in the design of structures throughout the system and, sometimes even, at the irrigation department or national level. He reasoned that natural conditions such as soil and climate are heterogeneous, especially in the case of large irrigation system, and may cause significant differences in the required irrigation service.

While acknowledging that simplicity might be the best option for some irrigation systems with respect to ease in operation and maintenance as suggested by Horst (1998), he expressed doubts that simple management solutions and techniques can also cope with the increasing complexity of water management when confronted with challenges posed by increasing complexity of cropping patterns, climate and multiple uses of water. He also questioned the concept of equity and uniformity in the conventional design, saying that it may no longer be an efficient management strategy. He suggested that the importance of equity should be re-addressed in terms of both quantity and quality of water.

He argued that a strengthening of the notion of service should be a key aspect of re-engineering for irrigation management. Such strengthening would have to include: (1) an upward service-oriented approach, starting from the level of the agreed service with the users and moving to the mobilisation of resources to achieve that service; (2) promoting an agreed cost-effective service by defining the level of service that meets both the demand and willingness of users to pay and by adopting water use indicators such as adequacy and efficiency to monitor effectiveness of irrigation service providers; (3) enlarging the concept of flexibility in water service to encompass the spatial variability of the service within an irrigation system; (4) defining and achieving the reliability requirements by users. For canal operations, the re-engineering approach should focus on an improved physical assessment of the infrastructure. In doing such an assessment, he proposed the use of sensitivity analysis to better understand canal behaviour under perturbations so that more relevant management strategies can be developed. He also called for a smart engineering approach to information systems for irrigation, advocating that a strategic and hierarchical approach to information pertinent to irrigation and the availability of water balance studies are important requirements to the design of cost-effective information systems. Further, he proposed that the service-operation interface be redesigned to include more appropriate allocation of operational effort targeting the most demanding areas.

5. ASSESSMENT OF IRRIGATION SYSTEMS

5.1 Approaches to analyze the hydraulic behaviour of irrigation systems

The conveyance and distribution of water in canal irrigation systems are governed by hydraulic laws, which either relate to steady state or unsteady state hydraulic conditions. Steady-state open flow hydraulics covers flows for which discharge is constant with time. It links the geometry, hydraulic properties and flow for different types of flow control structures and canals. In contrast, unsteady flow hydraulics considers discharge variation with time (Chow, 1959).

Steady state hydraulics is the more generally understood and developed topic. Steady-state flow conditions have been a widely used design assumption in most gated canal irrigation systems and are canal operation targets (Plusquellec *et al.*, 1994; Renault, 2001; Kouchakzadeh and Montazar, 2005; Renault *et al.*, 2007). However, they rarely happen in common irrigation practice as a canal operation upstream (e.g.: a change, or adjustment in gate setting) itself causes variations in discharge and water levels (or unsteady state conditions) along the canal network. It has been observed in many large canal irrigation systems that the time lag between a change upstream of the canal and its conversion at a downstream point of the canal can take a number of days. More often, new changes in flow conditions as a result of the succeeding operation are already beginning to be reflected downstream just when the flow conditions are converging after the preceding operational procedure (Renault *et al.*, 2007). Thus, the steady-state approach has limited practical significance for efficient management of canal irrigation systems, while it is not capable of analyzing the propagation of flow variations in discharge and water level along the canal networks (Renault, 2001; Kouchakzadeh and Montazar, 2005). Nevertheless, knowledge on steady-state flow properties of a canal system is relevant for purposes of measuring, monitoring and managing discharges at different locations along the canal network.

Unsteady flow condition is a norm in canal irrigation systems (Plusquellec *et al.*, 1994; Renault, 2001; Plusquellec, 2002; Renault *et al.*, 2007; Facon *et al.*, 2008). Therefore, the unsteady flow approach is suitable for analyzing the flow dynamics of irrigation distribution systems and canal operation (Renault, 2001; Renault *et al.*, 2007). Cognizant of the fact that perturbations or variations in discharges and water levels are permanent features of canal irrigation systems, consideration of unsteady state flow instead of assuming steady state has also been suggested for the design and layout of canal irrigation systems and for canal operation (Plusquellec *et al.*, 1994; Renault *et al.*, 2007; Renault, 2008). Several methods, including hydrodynamic models, based on Saint Venant differential equations have been proposed and used to describe unsteady flow conditions (Renault *et al.*, 2001). While an unsteady approach allows comprehensive analysis of hydraulic behaviour of canals, it is not simple (Renault, 2001; Renault, 2005; Kouchakzadeh and Montazar, 2005). Canal hydraulic models are seldom used by irrigation agencies due to their apparent complexity (Burt and Styles, 1998).

Canal hydraulic models are basically based on Saint-Venant equations developed by a Barré de Saint-Venant in 1871 (Litrico *et al.*, 2005). Saint-Venant equations are based on mass conservation law and momentum conservation law and are expressed as:

$$\frac{\partial A}{\partial t} + \frac{\partial Q}{\partial x} = 0 \tag{5.1}$$

$$\frac{\partial Q}{\partial t} + \frac{\partial Q^2/A}{\partial t} + gA\frac{\partial y}{\partial x} = gA(S_b - S_f) \tag{5.2}$$

where $A(x, t)$ is the wetted cross-sectional area (m^2), $Q(x, t)$ is the discharge (m^{3}s^{-1}) across section A, $Y(x, t)$ is the water depth (m), S_b is the bed slope, G is the gravitational acceleration (m s^{-2}), $S_f(x, t)$ is the friction slope.

Two boundary conditions are considered: $Q(0, t)$ and $Q(X, t)$ where X is the length of the canal. The initial conditions are denoted by $Q(x, 0)$ and $Y(x, 0)$. S_f is based on Manning-Strickler equation:

$$S_f = \frac{Q^2 n^2}{A^2 R^{4/3}} \tag{5.3}$$

where n is the Manning coefficient (s m$^{-1/3}$) and R is the hydraulic radius (m), defined by $R=A/P$ where P is the wetted perimeter.

Some of the hydraulic models that have been used for canal irrigation include MIKE 11 (Mishra *et al.*, 2001), SETRIC (Depeweg and Paudel, 2003), HEC-RAS (Shahrokhnia and Javan, 2005), CANALMAN (Kumar *et al.*, 2002; Khan and Ghumman, 2008; Ghumman *et al.*, 2010), ICSM (Ojo and Otieno, 2010) and SIC (Tariq and Latif, 2010). Other available hydraulic models include CANAL, CARIMA, DUFLOW, MODIS, USM, RootCanal, CanalCAD and Sobek. Evaluations for each of the first five models were made by Merkley and Rogers (1993), Holly and Parish (1993), Clemmens *et al.* (1993), Schuurmans (1993) and Rogers and Merkley (1993), respectively. CanalCAD and Sobek were evaluated by Clemmens *et al.* (2005). Merkley (2006) provides a description and technical reference of RootCanal.

In between the steady state and unsteady state approaches is the sensitivity analysis that focuses on the behaviour of canals under input changes. Rather than resorting to complexity of unsteady state flow hydraulics, this sensitivity approach considers two different steady states with corresponding slightly different value of inputs. The sensitivity of an irrigation structure is expressed as the ratio of the relative or absolute variation of outputs to the relative or absolute variation of the inputs. Depending on the function of the structure, the input and output are either water level or discharge. For a diversion structure or an offtake, the sensitivity indicator refers to the relative variation in discharge through the offtake ($\Delta q/q$) as a result of the variation in water level (ΔH) in the parent canal and is expressed as:

$$S_{offtake} = \frac{\Delta q / q}{\Delta H} \qquad (5.4)$$

Field measurements of the offtake sensitivity were done by generating a variation in the water level or head (ΔH) in the parent canal and measuring the corresponding variation of discharge (Δq) through the offtake. When records on water levels and discharges are available, the variations are obtained from the data and substituted to equation 5.4. For a rough estimate of the sensitivity indicator, the logarithm derivative of a generic equation of flow through the offtake will be used.

Generic flow equation: $q = cH^{\propto}$ $\qquad (5.5)$

Logarithm derivative: $\quad\quad\quad\quad \Delta q/q = \propto \frac{\Delta H}{H}$ (5.6)

Based on equation 5.4, equation 5.6 can be written as,

$$S_{offtake} = \frac{\propto}{H}$$ (5.7)

where H is the head exercised on the structure (water level upstream minus the water level downstream if the structure is submerged, or minus a level of reference taken as the crest level for an overshot structure, or the orifice axis for an undershot structure if the structure is not submerged). α is the exponent in the relevant hydraulic equation for flow; it is equal to 1.5 for overshot flow and 0.5 for undershot flow.

Offtake sensitivity can be used to asses discharge variations for different water levels in the main canal and to set water-level control requirements for a set water delivery target. Based on equation 5.4, the permissible variation in water level can be derived as:

$$\Delta H_{permissible} = \frac{\left(\Delta q/q\right)_{set}}{S_{offtake}}$$ (5.8)

A variation in diverted discharge through an offtake will have corresponding variation in discharge in the parent canal. Depending on the ratio of the discharge through the offtake to the discharge in the parent canal, such perturbation might be significant in the parent canal downstream of this diversion point. Thus, low sensitivity and high discharge offtake can have a large impact on perturbation along the main canal. The sensitivity indicator for conveyance expresses the relative variation in the main canal discharge as a function of the variation in water level (equation 5.9). While $S_{offtake}$ can be used to determine the impact of perturbations to the offtaking canal, $S_{conveyance}$ can be used to determine the impact of the fluctuations in the offtaking canal on the main system. If the offtake sensitivity is known, it is given by equation 5.7.

$$S_{conveyance} = \frac{\Delta Q/Q}{\Delta H}$$ (5.9)

With $\Delta q = \Delta Q$, equation 5.9 can be expressed as a function of the offtake sensitivity (equation 5.4).

$$S_{conveyance} = S_{offtake} \, q/Q$$ (5.10)

The water level control (cross-regulator) sensitivity indicator is expressed as the variation in water level or head (ΔH) resulting from a relative discharge variation ($\Delta Q/Q$) in the main canal. It is calculated using equation 5.11:

$$S_{regulator} = \frac{\Delta H}{\Delta Q/Q}$$ (5.11)

The methods to determine the sensitivity of cross-regulator are basically the same as those for the offtakes; the difference is the input and the output to be measured. Direct measurement is done by generating discharge variation in the main canal and measuring the resulting variation in the head upstream of the regulator.

As cited by Renault *et al.* (2007), studies on sensitivity analysis with emphasis on delivery structures were carried out as early as mid-20th century by Mahbub and Gulhati (1951) while further studies were conducted by Horst (1983), Albinson (1986), Ankum (1993) and Renault (1999). Recent studies (Renault, 2000; Kouchakzadeh and Montazar, 2005) involved development of hydraulic sensitivity indicators at reach and subsystem levels based on progressive aggregation of sensitivities. Indicators of sensitivity for each flow control structures are determined and aggregated at the reach level, then reach indicators are grouped at subsystem level.

At the canal reach and subsystem levels where a number of flow control structures might be interacting and influencing each other, it is the ratio of the rate of change in discharge from the outlet to the rate of change in discharge of the parent canal that are being assessed. It is termed as node sensitivity or hydraulic flexibility for ungated systems, and reach sensitivity for gated systems.

The node sensitivity (Hydraulic flexibility) indicator expresses the link between the relative variations in discharge (ΔQ) in the parent and dependent canals (Δq) at diversion or division nodes, and is calculated using equation 5.12. Hydraulic flexibility is especially well adapted to ungated systems that were developed on the principle of proportionality. Dividing both the discharge variation in the parent ($\Delta Q/Q$) and discharge variation in the dependent canal ($\Delta q/q$) by the water depth variation in the parent canal (ΔH) leads to an expression of hydraulic flexibility as the product of offtake sensitivity and water level regulator sensitivity (equation 5.13).

$$F = \frac{\Delta q/q}{\Delta Q/Q} \hspace{6cm} (5.12)$$

$$F = \left(\frac{\Delta q/q}{\Delta H}\right)\left(\frac{\Delta H}{\Delta Q/Q}\right) = S_{offtake} * S_{regulator} \hspace{3cm} (5.13)$$

Based on the generated discharge fluctuations in the offtaking canal, Horst (1998) categorized hydraulic flexibility indicators as follows:

• F < 1 (under proportional): a relative change in discharge in the parent canal generates a smaller relative change in the offtaking canal. Fluctuations are diminished in the offtaking canal;

• F = 1 (proportional): a relative change in discharge in the parent canal generates an equal relative change in the offtaking canal. Fluctuations are divided uniformly;

• F > 1 (hyper proportional): a relative change in discharge in the parent canal generates a larger relative change in the offtaking canal. Fluctuations are exacerbated in the offtaking canal.

In ungated irrigation systems developed on the principle of proportionality, the ideal value of the flexibility indicator is unity (F =1). In such cases, the discharge is divided proportionally over canals and a high level of equity can be achieved (Renault et al., 2007). In gated systems, the flexibility indicator approaches unity when the sensitivity indicators of the offtakes and the cross-regulators are inverse (equation 5.13). However in such systems, a number of strategies for managing perturbations is possible through gate adjustments, thus, proportional distribution of perturbations is not necessary a target of canal operation.

In gated systems, sensitivity analysis at the reach is important as flow control structures are interacting and influencing one another and conveying their behaviour downstream or upstream. Understanding the final behaviour of a reach requires aggregation of sensitivity indicators of several offtakes under a cross-regulator influence. The equation by Renault *et al.* (2000) will be used for computing for lumped sensitivity indicators at reach levels:

- sensitivity indicator of water depth (S_{RH}):

$$S_{RH} = \frac{1}{\sum_{i=1}^{n+1}(q_i/Q_{in})m_i S_1(i)} \tag{5.14}$$

- sensitivity indicator for conveyance (S_{RC}), expressed as the ratio of the perturbations leaving and entering the system:

$$S_{RC} = \frac{S_{RH}}{S_G}\frac{Q_{out}}{Q_{in}}, (S_{RC} \leq 1) \tag{5.15}$$

- reach sensitivity of delivery (S_{RD}), the ratio of the delivery perturbation to the inflow perturbation:

$$S_{RD} = \frac{\Delta Q_{del}}{\Delta Q_{in}} = 1 - S_{RC}, (S_{RD} \leq 1) \tag{5.16}$$

where q_i is the discharge through the offtake i, m_i is a coefficient of water depth variation within a backwater curve at any location i along a reach and at a downstream cross regulator, $S_i(i)$ is the sensitivity of the delivery for offtake i, ΔQ_{in} is the variation of discharge entering a canal reach, ΔQ_{del} is the variation of discharge delivered within a canal reach, ΔQ_{out} is the variation of discharge leaving a canal reach, and S_G is the sensitivity indicator for a downstream cross-regulator.

5.2 Overview of the performance assessment of irrigation systems

Developing irrigation facilities has been a key strategy of many countries to increase crop production. However, many studies reveal that the target irrigated area and crop production were not achieved in most areas served by canal irrigation systems. The evidently poor economic returns from irrigation investments and the increasing demand for crop production from irrigated agriculture, while its share of water is predicted to decline, has underscored the need to improve the performance of most canal irrigation systems. The emphasis on improving the performance has focused the attention of irrigation professionals on the approaches to investigate the causes of such mediocre performance and on the process of identifying improvement measures (Chambers and Carruthers, 1986; Bruscoli *et al.*, 2001).

Irrigation performance relates to the extent or level an irrigation system has met its set objectives (Murray-Rust and Snellen, 1993; Bos *et al.*, 2005; Malano and Hofwegen, 2006). Based on the nature of the irrigation system's objectives, two types of irrigation performance were defined by Murray-Rust and Snellen (1993): operational performance and strategic performance. The former measures the extent to which target levels of service are being met at any period in time through routine implementation of operational procedures. The latter is concerned with a longer-time activity that assesses the extent and the level of efficiency that the broader set of objectives is being achieved.

Studies on performance assessment of irrigation schemes have gained momentum since late 1980s (Gorantiwar and Smout, 2005). Performance assessment in irrigation is defined as the systematic observation, documentation and interpretation of activities related to irrigated agriculture with the objective of continuous improvement (Bos *et al.*, 2005; Molden *et al.*, 2007). It is the measurement of the degree of achievement of the strategic and operational objectives set out in the strategic planning (Malano and Hofwegen, 2006). Bos *et al.* (2005) and Molden *et al.* (2007) listed the following application of performance assessment:

- operational performance assessment to determine how the operational processes are performing;
- strategic performance assessment to understand how a system or organization is performing and using available resources;

- diagnostic performance assessment to understand the causes of low or high performance, and as a prelude to design and implementation of interventions for system improvement and rehabilitation;

- comparative performance assessment and benchmarking to compare performance to set appropriate benchmark standards and identify processes that lead to higher performance;

- assessing impact of irrigation to understand whether investments have paid off and understand the determinants of high or low productivity;

- supporting adaptive management to respond to changing needs and environment.

5.3 General framework for assessing irrigation performance

Building on the previous works of several authors, Bos *et al.* (2005) formulated a framework to define and guide the performance assessment task. The framework (Figure 5.1) is based on a series of questions on: (1) the purpose and strategy of performance assessment – to whom it is for, from whose perspective it is undertaken, who will carry it out, what is the type and extent of it; (2) the design and planning of the performance which include selecting criteria, performance indicators and data to be collected; (3) the implementation with data being collected, processed and analyzed; (4) the course of action on the results of data analysis.

The purpose of the assessment is closely related to who the assessment is for, which might be the government, funding agencies, irrigation service providers, irrigation system managers, farmers or research organizations. The assessment may be undertaken on behalf of a group of stakeholders but may be looking at performance assessment from the standpoint of another group of stakeholders. An example is an assessment study on the impact of irrigation system performance on farmer livelihoods to be carried out by a research institute hired by the government.

Different types of performance assessment will require different types of organizations or individuals to conduct the assessment as organizations or individuals differ in their capabilities in carrying out the assessment. The purpose of the assessment also defines the type of assessment, which can be operational, accountability, intervention, sustainability (Small and Svendsen, 1992) and diagnostic analysis (Bos *et al.*, 2005).

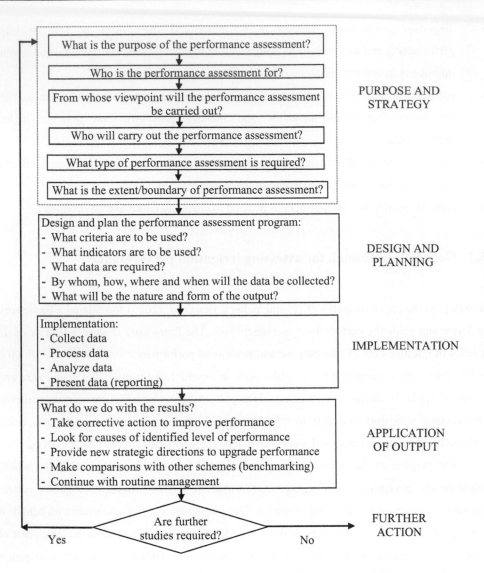

Figure 5.1. Framework for performance assessment of irrigation and drainage schemes

(Bos et al., 2005)

Operational performance assessment is defined as relating to the day-to-day, season-to-season monitoring and evaluation of system performance. Accountability performance assessment is undertaken to assess the performance of those responsible for managing the system. Intervention assessment is employed to study and enhance the performance of the system. Sustainability assessment looks at the longer resource use and impacts. Diagnostic analysis is carried out to track down the causes of performance so that improvements can be made or performance levels be sustained.

The performance assessment can either relate to only one system (internal analysis) or comparison between systems (external analysis). Benchmarking of irrigation systems is a form of external performance assessment. The two primary boundaries of performance assessment relate to spatial and temporal dimensions. Spatial refers to the area or number of schemes covered, that is, whether the assessment is limited to a particular canal within a system, to one system, or to several systems. Temporal refers to the duration of the assessment. The system approach advocated by Small and Svendsen (1992) is an example of a boundary and extent of performance assessment.

Bos *et al.* (2005) noticed that the terms performance criteria, performance measures and performance indicators have been used by different authors to mean different though closely related things. They proposed to equate the term criteria to the stated and/or alternative objectives of the irrigation system, against which performance is to be assessed. Criteria can be measured using performance indicators, which identify data needed for the performance assessment. Examples of performance criteria, which is also called as performance measures (Gorantiwar and Smout, 2005) and system characteristics (Schultz and De Wrachien, 2002), include efficiency, reliability, productivity, adequacy and sustainability, among others. Performance indicators are measurable variables that quantitatively describe the condition or performance in respect to the objectives established for the system. A number of papers have dealt with formulation and/or in-depth discussion on performance indicators (e.g.: Burt, 2001; Bos *et al.*, 2005; Gorantiwar and Smout, 2005). A comprehensive list of performance criteria, performance indicators and their data requirements are summarized by a number of authors (e.g.: Gorantiwar and Smout, 2005; Bos *et al.*, 2005).

5.3.1   Diagnostic assessment

The purpose of the assessment defines the nature of the performance assessment procedure that would be appropriate. A sound diagnosis of the current performance is often the most important phase in the modernisation process (Renault *et al.*, 2007), hence, a performance assessment procedure of diagnostic nature is given an emphasis in this review.

Diagnostic assessments are carried out to gain an understanding on reasons for the observed levels of performance, to find out the nature or causes of the problems, to know the constraints, to identify intervention measures for system improvement, or to learn more about successes and failures of irrigation design and management (Bos *et al.*, 2005; Molden *et al.*, 2007). It complements routine operational performance monitoring and evaluation of irrigation systems (Figure 5.2). It is undertaken when difficult operational problems are identified, when target performance is not met, or when stakeholders are not satisfied with the existing levels of performance achieved and desire a change.

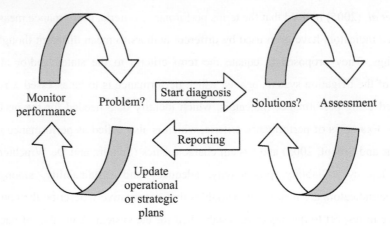

Figure 5.2. Relationship between the normal operational and strategic management cycle in irrigation (Bos et al., 2005)

There are two common approaches of diagnosing irrigation performance. The first approach involves collecting as much information as possible about the system and explaining the system performance through data analysis, while the second approach focuses on and tracing key cause-effect relationships (Bos *et al.*, 2005). The more cost-effective is the second approach, which takes a broad view of the system to identify possible explanations for the problem and offers explanations of reasons for the current level of performance after successive approximation. The basic steps for this approach are shown in Figure 5.3.

A system overview involves profiling of the system from macro perspective using secondary information on key descriptors of the irrigation system (climate, water supply and distribution, command area, crops, institutional framework, etc) in order to develop an initial explanation or hypothesis of cause-effect relationships. In a detailed assessment, more data can be collected and analyzed or certain parts of the system can be studied in more detail to reformulate and re-test the hypothesis (i.e., causes of problems) until it is validated, or supported by results of analysis. After finding the causes of the problem, recommendations are formulated and reported to concerned parties. Recommended actions may include adjustments in operations or in target values of indicators or formulation of new performance indicators.

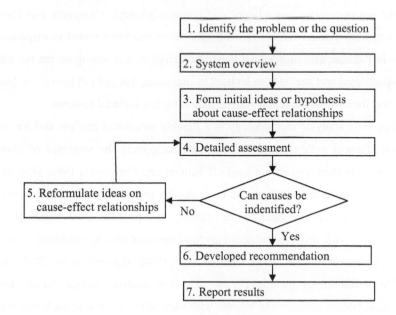

Figure 5.3. Steps to diagnostic assessment (Bos et al., 2005)

The methodologies applicable to diagnostic assessment include diagnostic analysis methodology, rapid appraisal, participatory rural appraisal and diagnostic tree. Examples of the more specific diagnostic assessment techniques include remote sensing, water balance and accounting and questionnaire surveys (Bos *et al.*, 2005). Diagnostic analysis and rapid appraisal as methodologies for assessing and understanding the performance of irrigation systems has been discussed in a number of papers (Chambers and Carruthers, 1986; Oad *et al.*, 1988; Derick *et al.*, 2000; Bos *et al.*, 2005). The authors traced back the roots of diagnostic analysis methodology from a joint on-farm water management research project by Pakistan government and Colorado State University in the early 1970s. In this pioneering work (Clyma *et al.*, 1977), a diagnostic analysis phase is one of the three phases of a so-called research development process (Figure 5.4). It involves an interdisciplinary study aimed at understanding the performance of the irrigation system and identifying the most important problems of the system. The general sequence of activities in a diagnostic study includes: defining preliminary objectives; reconnaissance (background information and preliminary surveys); revision of objectives and plans; detailed studies; interdisciplinary analysis and synthesis; and report writing (disciplinary and interdisciplinary). Chambers and Carruthers (1986) noted that the manual written up on diagnostic methodology used in irrigation water management in Pakistan and succeeding diagnostic activities was strong on the full checklist of data to be collected and methods to be used in diagnosis, but did not provide a diagnostic analysis method for a whole irrigation system, only for a few isolated systems.

The diagnostic analysis methodology is a flexibly structured process and has evolved over the years. Pioneer authors have defined more completely the sequence of steps for a diagnostic analysis in their succeeding works (Chamber and Carruthers, 1986; Dedrick *et al.*, 2000). Other contributors in the diagnostic methodology presented analytical frameworks or modified diagnostic processes based on the original concept (Oad *et al.*, 1988; Bos *et al.*, 2005). Further, diagnostic analysis methodology has been used also in combination with other methods (e.g., Hales and Burton, 2000; Dedrick *et al.*, 2000; Bruscoli *et al.*, 2001). The key concepts of the methodology include: (1) use of system analysis, which characterizes how various irrigation system components interact with each other; (2) use of an interdisciplinary team to carry out the assessment; (3) action research on developing data and knowledge that can be used to implement organization change; (4) users involvement throughout the process.

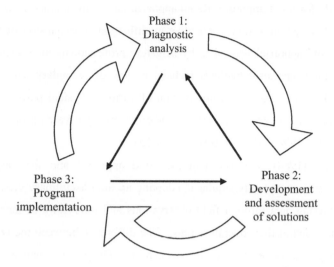

Figure 5.4. Research-development process (as cited by Oad et al., 1988)

The rapid appraisal method is a quick but systematic way to gather data on system performance. It falls on a continuum between very informal methods, such as casual talks or hurried site visits, and highly formal methods such as census, structured surveys, or field measurements (USAID, 1996). It is typically used in the initial steps of performing diagnostic analysis as its results, an overview of the irrigation system, can shed light in formulating an initial hypothesis on likely problems and in making decisions on promising courses of actions (Bos *et al.*, 2005).

The initial development of the rapid appraisal method can be traced to a research work of Bottrall commissioned by the World Bank on larger irrigation projects in Pakistan, India, Indonesia and Taiwan in the late 1970s (Chambers and Carruthers, 1986). The approach used by Bottrall and his two co-workers (an engineer and a local researcher) in each case study of a large irrigation system includes: (1) 1–3 weeks gathering information in the study area through interviews with farmers, cross-checking staff answers, inspecting records, spot checks; (2) 1–2 week general orientation including discussion with planners and administrators at the national level and brief visits to other systems for comparison. Oad *et al.*, 1988 cited other early works (De los Reyes, 1980; Chamber, 1983; Yoder and Martin, 1983; Potten, 1985; Latios, 1986) that contributed in the evolution of the concepts, methods, and checklist of data for rapid appraisal. Rapid appraisal has evolved since it was first employed in a number of irrigation systems. Adaptive modification or expansion of the concept and procedure of rapid appraisal was taken by irrigation professionals to suit specific situations. The common rapid appraisal methods include: review of secondary data; interviews with individuals and groups; and direct observation of various physical parts of the system and water supply conditions; and mini-surveys gathering quantitative data on narrowly focused questions (USAID, 1996 and 2010; Bos *et al.*, 2005).

Oad *et al.* (1988) remarked that the rapid appraisal and the diagnostic analysis methodologies are conceptually similar in adopting an interdisciplinary view of an irrigation system, in gathering data through field observations and interviews of farmers and staff of irrigation agency and in their overall purpose. The difference between the two approaches in diagnosing irrigation systems is in their emphases. Diagnostic analysis methodology is normally longer, more involved study with highly structured procedure or sequence of field activities based on more specifically defined concepts. Rapid appraisal gives more emphasis on the procedures of collecting and analyzing key information on the present condition and indicators of performance of the irrigation system in order to identify the likely cause of problems within relatively limited time.

As cited by Bos *et al.* (2005), participatory rural appraisal is a family of approaches and methods to enable local people to share, enhance and analyze their knowledge of life and conditions and to plan and act (Chambers, 1994). The local communities participate in the research by developing sketches and maps, transects showing resource use patterns, seasonal calendars, trend analysis and daily activity profiles. The participatory rural appraisal aims to tap local knowledge about the operation and performance of irrigation systems. It can be used as a complement to other tools when assessing performance, identifying problems and developing interventions. It is ideally suited for developing and improving service arrangements between the irrigation providers and users.

Performance of irrigation systems is determined by many factors such as technical design of the physical structures, quality of construction, system operation and maintenance, water supply conditions, agro-climatic factors, farmers' preference and farming practices, and institutional factors, among others. One or combinations of these factors are possible causes of a typical irrigation problem. Further, one cause might have originated from one or more causes. Diagnosis is done to sort out and determine which cause is leading to the observed field problem. Tracing the possible causes of the observed problem can be structured into a hierarchical diagnostic tree, where any given problem can be both a cause and an effect (Figure 5.5), and can led, eventually, to the identification of the root cause of the observed field problem (Burton *et al.*, 1999; Bos *et al.*, 2005). An example of the application of a diagnostic tree, also known as problem tree, in diagnosing an irrigation system is demonstrated in Bruscoli *et al.* (2001).

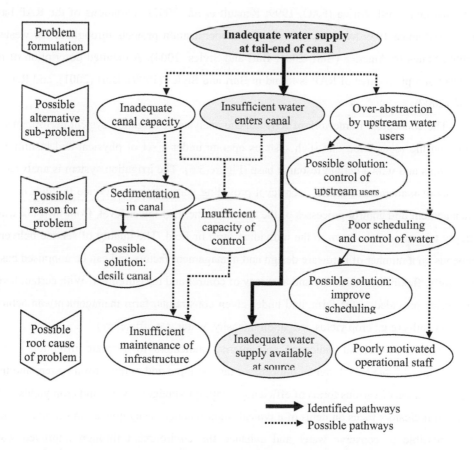

Figure 5.5. Structuring the problem and likely causes into a diagnostic tree

(Bos et al., 2005)

5.3.2 Rapid Appraisal Procedure

In the mid-1990s, a methodology to assess the impact of introduction of modern control and management practices on irrigation system performance was developed by the Irrigation Training and Research Centre (ITRC) at the California Polytechnic State University for a research program. This methodology called Rapid Appraisal Procedure (RAP) has been used since then by the World Bank and FAO for diagnosing irrigation systems in Asia, Latin America and North Africa (FAO, 1999; Renault *et al.*, 2007). Variations of the RAP have been used since 1989 by ITRC on irrigation modernisation projects throughout the western United States of America (Burt, 2001; Burt and Styles, 2004). A detailed description of the concept and procedures of RAP is found in Burt and Styles (1999), Burt (2001), and Renault *et al.* (2007).

The conceptual framework of RAP for the analysis of irrigation system performance is based on the recognition that such systems operate under a set of physical and institutional constraints and with a certain resource base (Figure 5.6). The irrigation system is analyzed as a series of management levels, each level providing water delivery service through internal management and control processes of the system to the next lower level, from the bulk water supply to main canals down to the individual farm or field. The quality of service delivered depends on a number of hardware design and management factors and can be appraised based on equity, flexibility, reliability and accuracy of control and measurement. With certain levels of service provided to the farm, and under given constraints, farm management can achieve certain results (e.g.: crop yields, irrigation intensity, water use efficiency).

RAP uses external indicators and internal indicators to evaluate irrigation system performance (Table 5.1). External indicators examine input and outputs for a system and they are expressions of various forms of efficiency relating to budgets, water and crop yields. RAP external indicators focus on items of a typical water balance. They give an indication whether it is possible to conserve water and enhance the environment through improved water management. Meanwhile internal indicators examine the processes and hardware used within the project. They quantitatively assess the project management and institutional setup, system operation, hardware used and water delivery service provided at all canal levels with the assigned values ranging from 0 (indicating least desirable) to 4 (denoting most desirable).

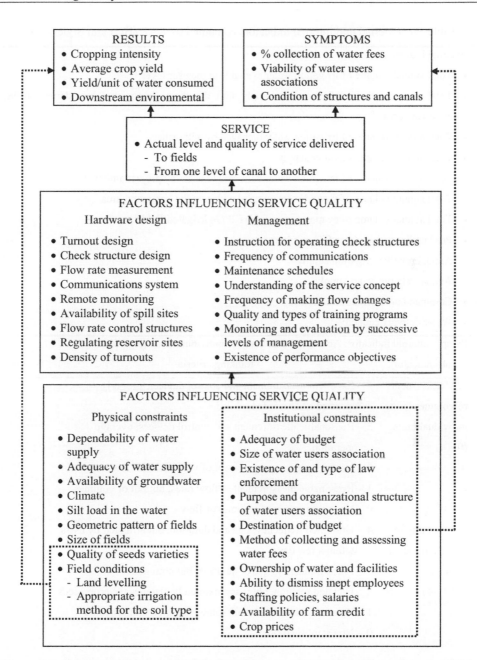

Figure 5.6. Conceptual framework of the Rapid Appraisal Procedure (Renault et al., 2007)

Table 5.1. External and internal indicators computed by RAP (Burt and Styles, 2004)

External indicators:
- Total annual volume of irrigation water available at the user level
- Total annual volume of irrigation supply into the three-dimensional boundaries of the command area
- Total annual volume of irrigation water managed by authorities
- Total annual volume of water supply
- Total annual volume of irrigation water delivered to users by project authorities
- Total annual volume of groundwater pumped within/to the command area
- Total annual volume of evapotranspiration (ET_c) in irrigated fields
- Peak net irrigation water ET_c requirements
- Annual relative irrigation supply
- Annual relative water supply
- Command area irrigation efficiency
- Water delivery capacity

Sample internal indicator: Actual water delivery to individual units (e.g., farm)

Sub-indicator	Ranking criteria	Weight[1]
1. Measurement of volumes to the individual units (0 – 4)	4 – Excellent measurement and control devices, properly operated and recorded	1
	3 – Reasonable measurement and control devices, average operation	
	2 – Useful but poor measurement of volumes and flow rates	
	1 – Reasonable measurement of flow rates, but not of volumes	
	0 – No measurement of volumes or flow	
2. Flexibility to the individual units (0 – 4)	4 – Unlimited frequency, rate and duration, but arranged by users within a few days	2
	3 – Fixed frequency, rate or duration, but arranged	
	2 – Dictated rotation, but it approximately matches the crop needs	
	1 – Rotation deliveries, but on somewhat uncertain schedule	
	0 – No established rules	

Table 5.1. *Continued*

Sample internal indicator: Actual water delivery to individual units (e.g., farm)		
Sub-indicator	Ranking criteria	Weight[1]
3. Reliability to the individual units (0 – 4)	4 – Water always arrives with the frequency, rate and duration promised. Volume is known.	4
	3 – Very reliable in rate and duration, but occasionally there are a few days of delay. Volume is known	
	2 – Water arrives about when it is needed and in the correct amounts. Volume is unknown.	
	1 – Volume is unknown, and deliveries are fairly unreliable, but less than 50% of the time.	
	0 – Unreliable frequency, rate, duration , more than 50% of the time, and volume delivered is unknown	
4. Apparent equity to individual units (0 – 4)	4 – All fields throughout the project and within tertiary units receive the same type of water delivery service.	4
	3 – Areas of the project receive the same amounts of water, but within an area the service is somewhat inequitable.	
	2 – Areas of the project receive somewhat different amounts (unintentionally), but within an area it is equitable.	
	1 – There are medium inequities both between areas and within areas.	
	0 – There are differences of more than 50% throughout the project on a fairly widespread basis.	

RAP makes use of a computer spreadsheet (Excel), which contains a range of questions related to water supply, personnel management, canal structures and level of water delivery service throughout the irrigation project that the evaluator must answer in a standardized format. Based on the data and information input on these worksheets, the internal and external indicators are computed automatically. Further, external and internal indicators provide a baseline of information for comparison against future performance after modernisation; benchmarking for comparison against other irrigation projects; and a basis for making specific recommendations for modernisation and improvement of water delivery service.

5.3.3   The MASSCOTE approach

The Mapping System and Services for Canal Operation Techniques (MASSCOTE) is a methodology for diagnosing canal irrigation system performance and analyzing modernisation of canal operation in order to improve the water delivery service and cost-effectiveness of canal operation (Renault *et al.*, 2007). It has been developed, and launched, by the FAO in response to the findings of its studies that highlighted inadequate attention to canal operation as among the major causes of underperformance of many canal irrigation systems.

MASSCOTE is a stepwise procedure of developing a modernisation plan starting from establishing baseline information on the state and performance of an irrigation system, capacity and sensitivity of the system, perturbation, water networks (flow route) and O&M cost to formulation of a vision of service-oriented management and modernisation of canal operation (Figure 5.7). Mapping the vision involves delimiting spatially manageable subunits based on demand for water delivery service and operation, defining the strategy for service and operation for each unit, and integration of service-oriented management options for the entire command area.

5.4 Conclusion

An assessment of initial conditions and performance of irrigation systems will provide useful inputs to better design and strategic planning of the physical and management improvements for the irrigation system. A diagnostic performance assessment using a combination of rapid appraisal, water budget and accounting tools (including in-situ measurements of seepage and percolation), and use of a hydraulic model is deemed as an appropriate assessment approach for NIS performance. Rapid appraisal is preferred over the standard diagnostic analysis for its time- and cost-effectiveness to give an overview of the current state and performance of irrigation systems and, at times, reasonable accuracy in tracing the root of problems, allowing for identification of intervention measures that can be taken (Chambers and Carruthers, 1986; Burt and Styles, 2004; Bos *et al.*, 2005; Renault *et al.*, 2007). Based on its results, initial hypothesis can be validated or rejected and a more specific field study can be developed.

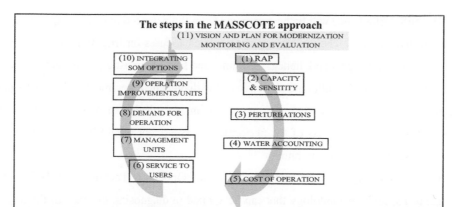

The steps in the MASSCOTE approach

(11) VISION AND PLAN FOR MODERNIZATION
MONITORING AND EVALUATION

(10) INTEGRATING SOM OPTIONS

(9) OPERATION IMPROVEMENTS/UNITS

(8) DEMAND FOR OPERATION

(7) MANAGEMENT UNITS

(6) SERVICE TO USERS

(1) RAP

(2) CAPACITY & SENSITITY

(3) PERTURBATIONS

(4) WATER ACCOUNTING

(5) COST OF OPERATION

Phase 1 – Mapping the baseline information
• performance. Initial rapid system diagnosis and performance assessment through rapid appraisal procedure (RAP), with three objectives: determine key indicators of the system to identify and prioritize modernization improvements; mobilize the energy of system managers and water users for modernization; and generate baseline assessment, against which progress can be measured.
• capacity and sensitivity of the system. Assessment of physical capacity of irrigation structures in performing the basic functions (conveyance, water level control, flow measurement, diversion, distribution, storage, safety, etc) and their sensitivities at key locations.
• perturbations. Analysis of the causes, magnitudes and frequency of perturbations to identify management or coping options.
• water network and water balance. Assessment of the hierarchical structure and main features of the irrigation and drainage networks, natural surface streams, groundwater and drainage system to know where and where all the inflow points to and outflow points from the service area occurs in terms of flow rates, volume and timing.
• cost of O&M. Disaggregation of cost associated with current operational techniques and cost analysis of options for various levels of services with current and with improved techniques.
Phase 2 – Mapping a vision of service-oriented management (SOM) and modernization of canal operation.
• service to the users. Assessment and cost analysis of the potential range of services to be provided to users as an initial step for crafting a preliminary vision of the irrigation scheme.
• management units. A subunit approach; partitioning of service area into manageable units on the basis on participatory management, spatial variations and requirements for water services, conjunctive water management, drainage conditions, among others.
• demand for operation. Assessment of the resources, opportunities and demand for canal operation; the higher the sensitivity, perturbations and service demanded, the higher the demand for canal operation.
• options for canal operation improvement. Identification of cost-effective options (service and economic feasibility) for improvements each management unit for water management, water control and canal operation.
• integration of SOM options. Improvement options for the subunits are finalized together with associated costs for every option, aggregated for the entire command area and checked for consistency within finalized improvement options at the main system level. A strategy for modernization is laid out with proposed achievements/ improvements.
• a consolidated vision and a plan for modernization and monitoring and evaluation (M&E). Consolidation of the vision for the irrigation scheme; finalizing a modernization strategy and progressive capacity development; selecting/choosing/deciding/phasing the options for improvements with the users; and developing M&E system.

Figure 5.7. The MASSCOTE framework (Renault et al., 2007)

The NIS performance will be best assessed from the perspective of the irrigation system design and management. The system approach, which focuses on irrigation system's inputs, process, outputs and impact and, hence, recommended for a diagnostic assessment (Bos *et al.*, 2005) will be adopted. The ultimate objective of the diagnostic assessment of NIS is to gather or generate relevant information or guidelines in the design and development of a modernisation plan. The results of the performance assessment would be useful to the NIA and to the research/scientific community.

The Mapping System and Services for Canal Operation Techniques (MASSCOTE) approach is a specific methodology that can be applied in diagnosing the state of the physical structure and analyzing canal operation modernisation. It conforms to the nature of the desired assessment approach described above, except that it does not include hydraulic simulation. A service-oriented management of improving water service delivery is central to the modernisation approach of MASSCOTE. As noted by Ertsen (2009), the list of the pertinent issues in irrigation within the context of service delivery is most complete in MASSCOTE documents.

Further the rapid appraisal procedure (RAP), a diagnostic tool that quantifies both the external and internal indicators of irrigation system performance, is embedded in MASSCOTE. It examines external inputs such as water supplies and outputs such as water destinations (e.g.: evapotranspiration, runoff). It provides a systematic examination of hardware and processes used to convey and distribute water to all levels within the irrigation system, from the source to the field. The RAP external indicators involve typical water balance parameters, thus, would give an idea whether or not it might be possible to conserve water through improved water management. Its internal indicators give a detailed perspective of how the system is operated and the water delivery service that is provided at all levels.

The use of both external and internal indicators is necessary for developing a relevant modernisation plan. External indicators and traditional benchmarking indicators are expressions of efficiency (ratio of output to input) and, by themselves, only indicate that things should be improved and provide only little guidance as to what must be done to accomplish such improvement. Internal indicators which examine the processes and hardware used within the system are necessary to identify intervention measures to improve the external indicators.

The RAP was originally developed by the Irrigation Training and Research Centre of California Polytechnic University and has been used by the FAO and the World Bank for appraising irrigation projects in Asia, Latin America and North Africa. Its use in irrigation evaluation and appraisal is recommended by the FAO because of its rapid nature, systematic procedures, and comprehensive approach as it covers the physical, management and institutional aspects of an irrigation system (Renault *et al.*, 2007).

In MASSCOTE, the analysis of the behaviour of irrigation structures is done through the assessment of their sensitivity: (1) for each type of main structure taken in isolation; (2) for a combination of associated structures; (3) at the reach and subsystem levels. The sensitivity of the flow control structures is defined as the ratio of a rate of change in output to the rate of change in the input. The input and output refer to either water level or discharge, depending on the function of the structure. The sensitivity analysis approach has also been proposed by a number of authors (Renault, 1999: Renault, 2001; Kouchakzadeh and Montazar, 2005) as an intermediate approach between the simple but limited steady state approach and the comprehensive but complex unsteady state approach. While hydraulic models can also be used, sensitivity analysis is considered a practical approach that makes more explicit the behaviour of irrigation structures (Renault *et al.*, 2007; Renault, 2008).

6. REHABILITATION OF NATIONAL IRRIGATION SYSTEMS

6.1 Overview on rehabilitation

The major irrigation development projects in the 1970s included construction of new irrigation structures as well as rehabilitation and improvement works of existing individual irrigation systems (Table 6.1, Figures 6.1 and 6.2, Annex C). These projects were mainly funded by foreign financial institutions, especially prior to 2000. Rehabilitation and improvement projects were carried out to either achieve 100% irrigation coverage of the service area; extension of the service areas, integration of the existing systems to form parts of the new, bigger irrigation system or multi-purpose water development projects or a combination of any of these objectives. The major projects focusing on rehabilitation and improvement of irrigation systems in the country started in the late 1970s. Among the first of these projects were the Laguna de Bay Irrigation Project, Second Laguna de Bay Irrigation Project and the first and second Irrigation Operations Support Projects implemented in Sta. Maria RIS, Balanac RIS and in both systems, respectively.

The main criticism on rehabilitation stemmed from the perceived frequent needs to rehabilitate the existing NIS and the insignificant increase in cropping intensities. Despite of continuing rehabilitation and improvement works, often combined with a notion of participatory irrigation management, the NIA data show that the areas irrigated by the NIS for the period 2000-2014 averaged 73 and 71% for the wet and dry seasons, respectively. The corresponding cropping intensity was 144%. David (2008, 2009) estimated even a lower dry season irrigation intensity in the order of 50-60% of the design irrigation service area.

The apparent norm of lower actual area served by NIS has been attributed to shortcomings either in the design, technology, system operation and maintenance or to a combination of these factors. David *et al.* (2012a) concluded that the low dry season irrigation intensities of the four canal irrigation systems in Ilocos Norte were mainly due to design shortcomings in the headworks. These shortcomings included underestimation of flood flows and sediment loads, inadequate provision for sediment control and underestimation of reservoir inflow and outflow hydrographs. The results of their complementary study on nine

NIS (David *et al.*, 2012b) showed that the lower percolation rates used during the design stage of the project resulted in the gross underestimation of irrigation water requirement and, hence, design canal discharge capacities. David (2003) and David *et al.* (2012a, 2012b) argued that the insignificant impact of rehabilitation efforts has been mainly due to the adopted approach of restoring the original physical structure without rectifying unrealistic design considerations.

Table 6.1. Completed and on-going major rehabilitation projects

Project name	Duration	Cost (10^6 US$)
1. Upper Pampanga River	1969–1976	34.0
2. Angat-Magat Integrated Agricultural Development	1973–1980	9.6
3. Tarlac Irrigation System	1974–1980	17.0
4. Magat River Multi-purpose Project 1	1975–1982	42.0
5. Chico River Irrigation Project	1976–1981	50.0
6. Jalaur River	1976–1982	15.0
7. National Irrigation System Improvement Project (NISIP)	1977–1985	76.0
8. Second Magat River Multi-purpose Project	1978–1983	150.0
9. Magat River Multi-purpose Project Stage II- Irrigation	1979–1983	21.0
10. NISIP 2	1978–1987	80.7
11. Second Laguna de Bay Irrigation Project	1981–1987	
12. Irrigation Operations Support Project (IOSP) 1	1988–1993	69.1
13. Irrigation Systems Improvement Project (ISIP)	1991–1996	36.3
14. IOSP 2	1993–2000	51.3
15. ISIP 2	1996–2006	34.78
16. Water Resources Development Program	1997–2005	58.5
17. Angat-Maasim El Niño Mitigation Measures	1998–1999	
18. Southern Philippines Irrigation Sector Project	1999–2011	80.9
19. Agno River Integrated Irrigation Project	2006–2013	89.1
20. Participatory Irrigation Development Project (PIDP)	2009–2024	413.6
21. Irrigation System Operation Efficiency Improvement Project	2010–2016	1.3
22. Irrigation Sector Rehabilitation and Improvement Project	2010–2016	JPY6,187.0
23. National Irrigation Sector Rehabilitation and Improvement Project (NISRIP)	2013–2016	PhP4,007.1

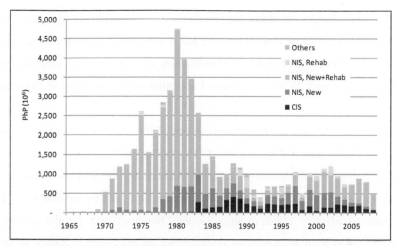

Figure 6.1. Distribution of public expenditures for irrigation investments for foreign assisted projects, by type of system and purpose, 1965-2008 at 1985 prices (data source: NIA)

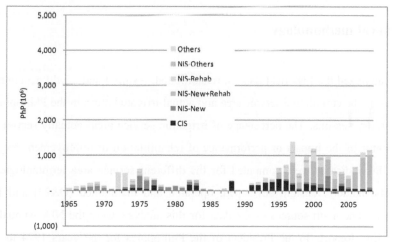

Figure 6.2. Distribution of public expenditures for irrigation investments for locally funded projects, by type of system and purpose, 1965-2008 at 1985 prices (data source: NIA)

Meanwhile, developments in the river basins and changes in weather patterns over time have modified the hydrographs of water supply sources for irrigation. Increased cropping intensity, crop diversification and other farming practices adopted by farmers to maximize profits require greater flexibility in irrigation water delivery service. How the design shortcomings and changes in water supply and irrigation demands are addressed in the rehabilitation or system improvement projects has a substantial bearing on their impacts. In other words, the success or failure of rehabilitation or improvement efforts considerably depends on how well such factors are assessed and adjustments are made in the rehabilitation designs.

In this chapter, the process, nature and impacts of the rehabilitation approach adopted in the NIS in the Philippines will be examined. The first objective of this examination was to assess the impact of rehabilitation in terms of maintaining the irrigation service area and closing the gap between the actual area irrigated and the design service area. The other objective was to identify effective improvement practices and pitfalls of the adopted rehabilitation approach so as to provide feedback for an improved irrigation modernisation approach, including enrichment of the available baseline information for planning irrigation system modernisation in the Philippines.

6.2   General methodology

The study analyzed the historical data on rehabilitated, restored and newly generated service areas vis-a-vis the cumulative service area and actual irrigated areas in the Philippines during the wet and dry seasons. The percentage of irrigation service areas actually served was used as an indicator of the impact or performance of rehabilitation or improvement projects. The values of this indicator were computed for the different service area terminologies used in NIA reports, namely: design, developed, operated and maintained (O&M) and firmed-up service areas. The main sources of the data for this analysis were the NIA Annual Reports, NIA Year End Reports to the President of the Philippines for the years 1974 to 2012 and records provided by the NIA Management Information Division.

Unlike the data on the developed and O&M service areas, the data on total area provided with irrigation facilities were not readily available. Historical data on these aggregate values of the design service areas of all irrigation systems that had been built would indicate the level of irrigation development efforts carried out so far. A data set on the aggregated NIS 'design' service area of the country was constructed to serve as one of the benchmarks of a trend analysis. It was constructed by using the reported O&M service area in

1979 as the starting value. The 'design' service areas for succeeding years were computed by adding the reported new NIS service areas generated each year thereafter. The year 1979 was selected because it was the year when different terminologies pertaining to irrigation service area approximate each other's reported values the closest.

The ratios of the irrigated area to the different service area terminologies were also computed for each of the irrigation systems under study, namely: Agos RIS, Balanac RIS and Sta. Maria RIS. The data on different service area terminologies were sourced from their corresponding Irrigation Management Office and system offices. The statistics on areas not irrigated by the three irrigation systems under study were verified with respective system staff and irrigators association (IA) leaders to ensure that data only included those farms where the irrigation systems failed to deliver water. Hence, non-irrigation of crops due to reasons other than water-related problems did not reflect negatively on the impact of rehabilitation efforts.

The program of works (POW) of the rehabilitation projects implemented for each of the systems since mid-1990s were obtained from the concerned NIA offices. Each individual item of work was categorized according to its nature or the physical component involved (Annex D). The analysis also included the three foreign-assisted projects implemented during the past three decades, namely: the Second Irrigation Operations Support Project and the Participatory Irrigation Development Project that covered the three systems between 1993–1999 and 2011–2012, respectively; and the Water Resources Development Project that was implemented in Balanac RIS from 1997–2001.

The key persons at different levels of the NIA offices (i.e., central, regional and field offices) were interviewed to solicit information and insights on the planning and implementation process of rehabilitation projects. A questionnaire was developed and used for the purpose of the interviews.

A walkthrough from the dam or main intake structure down to the canal network of each NIS was carried out to have a visual assessment of the operational status and physical condition of irrigation structures that had undergone rehabilitation and improvement. The operation and maintenance staff and farmers in the area were interviewed during the walkthrough. The interview questions mostly related to the previous rehabilitation projects, functional status of flow control structures, adequacy of water supply, water management, and system operation practices, quality of irrigation service and IMT, among others.

6.3   Definition of terminologies

A number of service area terminologies are used in NIA reports and documents. The definition for each terminology discussed in this thesis is given below.

- design service area is the area provided with irrigation facilities and from which the projections of benefits from the irrigation system were inferred from during the project feasibility and appraisal stages;

- O&M service area refers to the service area that has been turned over by the design and engineering unit of the NIA to the operations unit for actual operation;

- new generated area refers to the design service area of newly built irrigation systems during the period or year of reporting and new area covered by extending the original service areas of existing irrigation systems. It is an addition to the total design service area or developed service area reported for the previous period or year;

- developed NIS service areas include the O&M service areas as well as those newly generated areas that have not been turned over yet to the operations unit;

- rehabilitated area is a water-deficit irrigation service area where a reliable delivery of sufficient water is expected as a result of repairs or improvement in the physical structures influencing the flow of water to this service area;

- restored area is a part of the original service area that was deprived of irrigation water but can now be irrigated as result of rehabilitation or improvement works;

- firmed-up service area (FUSA) is defined as the estimated maximum area that can be served after unforeseen design and operational constraints as well as any over optimism on the estimation and delineation of the design service area are manifested by operational reality. It also accounts for any shift in land use from irrigated areas to non-irrigated areas or non-agricultural uses as well as any increase in service area as result of area restoration and extension works. FUSA is a system level data and was included in NIA Annual Reports starting 2012;

- programmed area refers to the area scheduled for irrigation in the forthcoming wet and dry cropping seasons. It is estimated by operations staff based on their assessment of water supply availability and expected O&M situations.

Both the developed and the O&M service areas also reflect changes in area resulting from land conversion to non-cultivated uses and permanent loss of service area due to insufficiency of water supply or inadequacy in irrigation facilities to deliver the water. Such changes account for the lower values of reported O&M service areas and developed service areas compared to the design service area.

Rehabilitation as defined by the NIA does not distinguish between system improvement and conventional rehabilitation, which re-establishes the physical design of the original structure. The same convention was adopted in the succeeding discussion for this Chapter.

6.4 Magnitude and impacts of rehabilitation or improvement projects

The total rehabilitated and restored service areas of NIS and communal irrigation systems (CIS) for the period from 1980 to 2014 totalled almost 4.8 million ha (Table 6.2). The average rehabilitation and restoration efforts in the 1980s, 1990s and 2000s covered 113,000 ha yr^{-1}, 112,000 ha yr^{-1} and 168,000 ha yr^{-1}, respectively. These efforts covered 195,000 ha yr^{-1} during the past five years (2010–2014).

The information in Table 6.2 points some questions with respect to the efficiency of planning and implementation of canal irrigation projects. During the same period, the new service area generated by constructing new NIS and CIS and expanding the areas of existing NIS and CIS beyond the original design area was almost 849,000 ha. Meanwhile, the combined net increase of developed irrigation service areas of these systems was about 563,000 ha. It is about 66% of the new service area provided with these irrigation facilities during the same period. This would mean that about 285,000 ha were lost to irrigation despite the significant rehabilitation efforts carried out to maintain the irrigation service area.

Almost 80% (3.7 million ha) of the combined NIS and CIS service areas rehabilitated and restored were NIS service areas (Figure 6.3). The average NIS rehabilitation and restoration efforts in the 1980s, 1990s and 2000s covered about 94,500 ha yr^{-1}, 94,900 ha yr^{-1} and 122,200 ha yr^{-1}, respectively. For the period 2010–2014 these efforts covered 141,200 ha yr^{-1}. This advantage of NIS over CIS in terms of magnitude of rehabilitation efforts resulted in a lesser NIS service area lost to irrigation of about 101,400 ha than that of CIS. The NIS service area lost was about 23% of the new NIS service area generated during that period.

Table 6.2. Cumulative, new generated and rehabilitated NIS and CIS areas (in 10^3 ha) nationwide

Year	Cumulative developed service area			Period	New area		Rehabilitated/restored area	
	NIS	CIS	Total		NIS	NIS + CIS	NIS	NIS + CIS
1979	475	549	1,024					
1980	472	577	1,049	1980	28	53	86	103
1985	568	704	1,272	1981-1985	142	217	315	393
1990	663	751	1,414	1986-1989	67	97	544	638
1995	652	474	1,126	1990-1995	17	61	470	520
2000	686	501	1,187	1996-1999	37	90	480	600
2005	696	543	1,239	2000-2005	56	104	670	868
2010	767	558	1,325	2006-2010	32	54	648	928
2014	822	766[1]	1,587	2011-2014	70	172	515	727
Increase	347	217	563	Total	448	849	3727	4778

[1]Included almost 1,700 ha of CIS assisted by other Government agency, a category introduced in 2012.

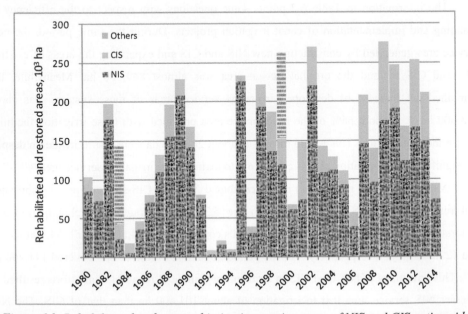

Figure 6.3. Rehabilitated and restored irrigation service areas of NIS and CIS nationwide

However, the actual area irrigated by NIS remained considerably smaller than any of the irrigation service area terminology used in NIA reports (Figure 6.4). The actual area served had the widest gap with the 'design' service area. Such gap is due to the fact that the design service area is a fixed value for a given NIS while the values for the other service area terminologies are adjusted periodically to account for lands that were permanently excised out of the service area due to conversion to non-irrigated uses or failure of irrigation service.

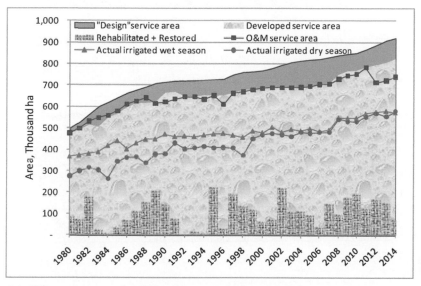

Figure 6.4. NIS service areas by different terminologies, rehabilitated and restored areas and actual irrigated areas by season

During the period 2000–2014, the actual area irrigated by NIS nationwide ranged from 57-65% of the "design" service area with average values of 62% and 61% for the wet season and dry season, respectively (Figure 6.5). The corresponding annual irrigation intensity of 123% was much less than the 160-200% levels projected during project planning stages. The percentages of irrigated area with respect to the design area were the lowest among their corresponding values in the case of the O&M design and developed service areas. This is because the design area is fixed for a given system while developed and O&M are adjusted from time to time. Such adjustments tend to distort the trends. The percentages of design area irrigated would indicate the efficiency of NIS development efforts.

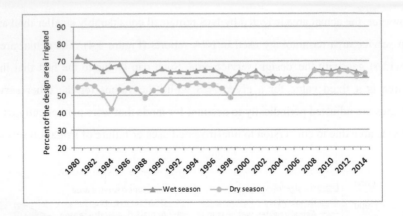

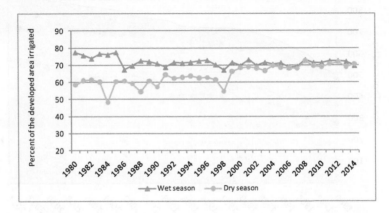

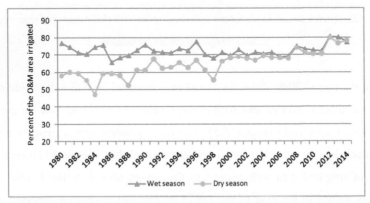

Figure 6.5. Percentages of NIS design, developed and O&M service areas irrigated, by

season

The percentages of the O&M service area irrigated ranged from 67-81% with an average of 73% for the wet season and 71% for the dry season. On average, they were 10% higher than those of the 'design' service area. However, the average annual irrigation intensity remained below the projected level at 144%. In terms of the developed service, the actual area irrigated ranged from 67-73%, with average values of 71 and 69% for the wet and dry seasons, and average cropping intensity of 140%.

It is noteworthy that the irrigation intensity associated with O&M and developed service areas remained below the level projected in the design despite the periodic reduction in the original service area. The trend of lower-than-projected irrigation intensities despite an average rehabilitation frequency of once in six to seven years is symptomatic for shortcomings in the design and management of irrigation systems. It also casts doubt on the effectiveness of the rehabilitation efforts.

In general, higher percentages of service area actually irrigated were achieved during the wet season than during the dry season for each service area terminology. This trend was more pronounced during 1980s to late 1990s. Smaller differences in actual area irrigated between the two cropping seasons can be observed starting 2000 as the values for the dry season gradually increased.

Establishing the reasons for the trend of increasing percentages of dry season irrigated area was beyond the scope of this study. However, it is reasonable to assume that much of the increase can be attributed to the growing popularity of shallow tubewell (STW) irrigation in NIS service areas that suffer from inadequacy of surface water supply. Beginning early 1990s, the STWs grew at a fast rate. The Department of Agriculture and the NIA launched programs promoting the installation of tubewells in NIS service areas. In the 2002 Census of Agriculture, the farm areas reporting presence of individual irrigation systems (primarily tubewells) totalled over a million ha.

The apparent trend of decreasing percentages of irrigated area during the wet season could be due to the following: a shift in cropping calendars among farmers with STW towards the less risky months from typhoons and floods; changing hydrographs; and the increasing incidence of flooding and siltation of canal irrigation service areas. The latter exacerbate damages to headworks and major canal systems and weaknesses in operation, management and flow control structures of irrigation systems to deal with extreme rain events.

6.5 Planning and implementation process of rehabilitation projects

Irrigation development and management in the country is the primary responsibility of the NIA. The NIA has three levels of offices in carrying out this mandate, namely: Central Office (CO), Regional Irrigation Office (RIO) and Irrigation Management Office (IMO). The CO issues guidelines and policies and exercises control over field operations nationwide. The RIO implement the ensuing plans, programs and policies of the NIA in their respective region, which is composed of a number of provinces. The IMO are responsible for the construction and rehabilitation of irrigation systems in their respective province or cluster of provinces, and for the implementation of the system O&M plans in collaboration with the farmer-irrigators. Foreign-assisted irrigation projects are implemented through the Project Management Offices (PMO) which are created and are co-terminus with their respective projects. There are 14 RIOs and 51 IMOs nationwide. The number of PMOs varies depending on the number of on-going foreign-assisted projects.

The planning of rehabilitation projects involves the CO, RIOs and IMOs. Each of these offices has an engineering and operations group that carries out the work involved in rehabilitation projects. At the CO level, locally funded rehabilitation projects are handled by its System Management Division while foreign-assisted and large or special rehabilitation projects are handled by its Engineering Department. The Planning and Programming Division of CO administers the budget allocation and prioritization process.

Proposals on rehabilitation and system improvement basically originate from an IMO (Figure 6.6). Depending on the physical component to be rehabilitated and the associated cost, POW and detailed engineering of proposed rehabilitation projects are carried out either solely by the IMO or in collaboration with the RIO or CO. Per Memorandum Circular 68 on Delegated Authorities of NIA officials issued in 2010, the IMOs and RIOs have basically the same delegated authorities concerning project engineering designs, plans and drawings which include the following: (1) farm-level facilities and parcellation maps of new national irrigation projects; (2) diversion dams lower than 2.5 m high and with drainage area of 10 km^2 or less; (3) canal and drainage structures costing less than PhP 25 million per structure; (4) main intake structures with service area of not more than 1,000 ha; (5) repair of river protection works/dikes; (6) repair of roads and appurtenant structures; (7) detailed design of diversion dams to be rehabilitated; (8) rehabilitation of canals and hydraulic structures. In the case of the NIS studied, the IMO prepares the engineering aspects of the proposals while engineers of the RIO review them or provide technical backstopping in the preparation.

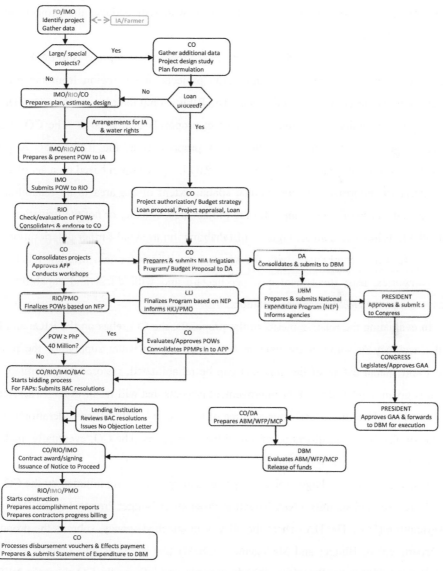

Figure 6.6. Schematic of the project formulation, development and implementation including budget execution (Adopted from NIA System Management Division)

Farmers' involvement usually includes identification of rehabilitation needs and preparation of the Memorandum of Agreement. In the case of the three NIS studies, identification and preliminary prioritization of rehabilitation needs is a joint undertaking of the system office of IMO and IA leaders. The irrigation system superintendent recommends the list of rehabilitation needs to the IMO head, who will endorse the projects to their Engineering Division.

Each IMO submits its priority list of proposed rehabilitation projects to the RIO, which in turn makes its list of priority rehabilitation projects of the region from the proposals forwarded by the different IMOs it oversees. The RIOs submit their lists to the CO when there is a call for submission of rehabilitation project proposals by the latter. The CO conducts deliberation of project proposals with project proponents at the RIO level. A priority allocation of funds for rehabilitation among the RIO is agreed upon based on a set of criteria. The criteria for prioritization may include spatial extent of the area to be rehabilitated or generated, adequacy of water supply for the project area, whether or not the irrigation system is situated in priority rice areas, urgency of rehabilitation to avoid further deterioration of the irrigation facilities, compliance with the set cost guidelines for the different rehabilitation work categories, least investment cost per hectare and willingness and capability of the farmers to manage O&M after rehabilitation.

In evaluating the relative merits of the proposals, the CO relies on the RIOs and IMOs for the accuracy of some of the data such as adequacy of water supply for the proposed project and spatial extent of the area that can be rehabilitated, restored or generated. More weight is given to rehabilitation or improvement projects that will result in larger service area maintained or added to the total irrigated area of the country. Political consideration is said to be a factor also in the prioritization of rehabilitation projects. The CO, eventually, makes the final selection of projects to be funded.

Together with other large projects or projects of special nature initiated by the CO itself, the CO prepares and submits a NIA Irrigation Program or Budget Proposal to the Department of Agriculture (DA). The DA, where the NIA is an attached agency, submits the proposal to the Department of Budget and Management (DBM) that prepares and submits a National Expenditure (NEP) to the President of the Philippines and informs the CO about the NEP. The President approves and submits the NEP to the Philippine Congress who legislates and approves the national budget (General Appropriation Act). While waiting for the enactment of the budget, the CO revises or finalizes its irrigation program based on the NEP and informs the concerned RIOs and PMOs, which in turn, finalize their POWs accordingly. The project proponents start the bidding process if the final POW worth less than PhP 40 million (US$ 920,000 as of October 2013). Otherwise, it is first submitted to CO for evaluation and approval. Once the national budget is passed by Congress and approved by the President, the DBM notifies the CO and DA and issues a notice of funds availability. Concerned project proponents will proceed with the awarding of contract and issuance of notice to proceed to the winning bidders for the project construction. For foreign-assisted projects, bids resolutions are submitted to the lending institution for approval prior to awarding and signing of a contract.

The implementation of rehabilitation projects is a responsibility of the IMO. It supervises the construction and accepts the completed work. The RIO or PMO coordinates with IMO in the preparation of financial and physical accomplishment reports and contractors' progress billing. The CO processes and effects payments and submits a statement of expenditure to DA and DBM.

At present, monitoring of project implementation and evaluation of the accomplishments of targets or objectives of completed projects is not part of the process of rehabilitation. Unlike in the case of foreign-assisted projects where review missions go to the field, higher offices of NIA rely on reports from their implementing offices regarding project completion. Impact evaluation after full project development stage is also not part of the rehabilitation program.

The benefits of involving the IAs in the planning and implementation process were well recognized by the NIA staff interviewed. The involvement of IAs is basically limited to identification of rehabilitation needs.

Though there is recognition of water adequacy as an important criterion in planning irrigation development and rehabilitation work, thorough technical assessments of water supply availability are not among the priority undertakings currently pursued by NIA. In the case of the three NIS studied measurement of river flow at diversion points has not been done for the past three decades. While water diverted for irrigation is being measured in Sta. Maria RIS at the initiative of its IA, it has not been measured in Agos RIS and Balanac RIS. Engineers and technical staff of IMO based their estimates of water adequacy for a proposed project on their field experience or observation in the past. The RIO relies on the IMO for the correctness of such information. Without river flow gauging activities in most NIS, the emphasis on water adequacy criterion would be reduced to just a lip service.

It was gathered from the interviews that there was a general tendency of the project proponents to focus their system improvement plans to rehabilitation works with apparent large target area in order to bring the cost per hectare within the cost guidelines set by the CO. The least investment cost criterion combined with the emphasis on total service expansion might be counterproductive as these would screen out rehabilitation or improvement works with relatively high initial investment even though they are more cost-effective in the long run.

6.6   Nature and impact of rehabilitation: a system level study

6.6.1   Magnitude of rehabilitation

The magnitude of rehabilitation efforts carried out in Balanac RIS, Agos RIS and Sta. Maria RIS during the period from 2000 to 2014 were at least 780 ha yr^{-1}, 440 ha yr^{-1} and 330 ha yr^{-1}, respectively (Table 6.3). These rates accounted for about 76% of FUSA of Balanac RIS and 34% of FUSA of Agos RIS and Sta. Maria RIS.

6.6.2   Nature of rehabilitation

Lining of canals was the most frequent and most invested work, accounting for about 40-60% of the total rehabilitation expenses for each NIS (Figures 6.7 and 6.8). Canal lining materials used included class B concrete mix, concrete hollow blocks and grouted riprap. The first two types of lining materials were the most prevalently used in canal lining works for Sta. Maria RIS, while grouted riprap was most common in Balanac RIS. In most cases, the canal lining works were carried out on earthen canals of the three NIS. The few exceptions were the canal re-lining works in Balanac RIS.

Headworks, roads and canal structures were among the next three most invested projects in the three NIS. They each comprised less than 10% of the total investment for respective systems with the exception of the dams of Balanac RIS and Sta. Maria RIS and the road in Sta. Maria RIS, which accounted for 39%, 12% and 16%, respectively. Headwork rehabilitation and improvements in Balanac RIS involved construction of an upstream protection wall and additional length of the crest and apron of the dam, provision of mechanical lifting mechanism for sluice gates, installation of safety railings and repairs of the sluice and main intake gates. In Sta. Maria RIS and Agos RIS, headwork rehabilitation and improvements included construction of approach protection walls or dikes, river channelling and repairs of main intake gates. They also involved brush damming and lengthening of dam aprons for Agos RIS and Sta. Maria RIS, respectively.

Table 6.3. New service area generated and rehabilitated or restored service areas (in ha)

Year	Agos RIS			Balanac RIS			Sta. Maria RIS		
	Rehab	Rest	New	Rehab	Rest	New	Rehab	Rest	New
2000	-	-	-	1,056	-	-	1,972	-	-
2001	-	-	-	1,056	-	-	-	-	-
2002	-	-	-	594	-	-	-	-	-
2003	1,310	-	-	-	-	-	800	-	-
2004	-	-	-	-	-	-	300	-	-
2005	1,700	-	-	1,530	-	-	540	-	-
2006	155	10	-	940	-	-	-	-	-
2007	109	8	-	730	50	-	-	-	-
2008	445	10	-	525	-	-	100	106	-
2009	230	73	-	80	-	-	187	30	-
2010	417	109	-	2,180	-	-	222	214	-
2011	1,254	189	30	370	50	-	574	42	-
2012	115	17	-	700	13	-	-	-	-
2013	850	10	-	70	1,000	-	-	-	-
2014	-	-	-	430	-	20	254	-	-
2015	30	-	-	1,090	-	-	-	-	-
Total	6,615	426	30	11,351	1,113	20	4,948	392	0

Canal structure works in Balanac RIS and Agos RIS primarily involved construction of proportional and crested weirs, flumes and siphons and reduction of turnout sizes. They primarily involved installation of end check gates and vertical gates, and pipe turnouts in Sta. Maria RIS. End check gates were also installed in Balanac RIS. End check gates were used to re-capture drainage water and divert additional water from nearby creeks.

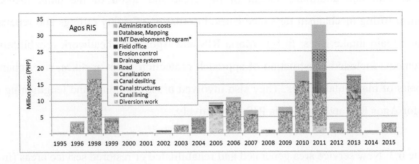

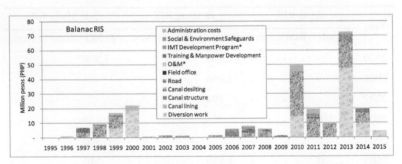

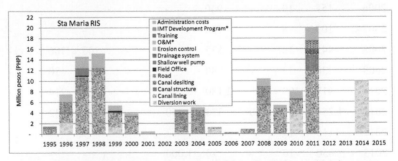

IMT - Irrigation Management Transfer; O&M - Operation and Maintenance

Figure 6.7. Rehabilitation/improvement cost, by component for the sample NIS

The rehabilitation and improvement projects carried out for the three NIS were mainly aimed at maintaining their respective service areas as implied in each project's reported accomplishments. They were carried out with an end view of enabling the physical system to irrigate the whole service area. Though commonly referred to as rehabilitation, most of such projects are mainly improvement works as they involved provision or construction of physical features that were not part of the original physical structures. In a way they were interventions to partly rectify the perceived design shortcomings and to adapt to changing hydrological regimes.

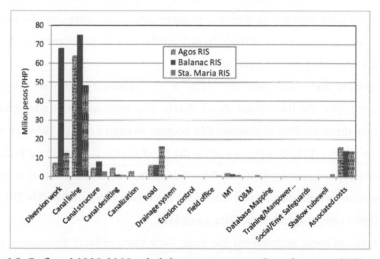

Figure 6.8. Deflated 1995-2015 rehabilitation expenses (based on year 2000 price) by component

The provision of concrete lining on earthen canals and the installation of check gates to reuse irrigation water or to divert additional water from creeks crossing the service area were interventions aimed at decreasing water conveyance losses and augmenting water supply, respectively. Systems officials and farmers deemed canal lining as means of improving conveyance efficiency and shortening the travel time of irrigation water. However, assessment of percolation and seepage rates, water supply adequacy and canal discharge capacity were not among the factors considered in canal lining project proposals.

The proportional weirs were constructed to replace the dilapidated, original spindle type vertical gates installed at major branching points of Balanac RIS, to simplify the system operations to distribute water. However, they were vandalized with significant length of their crests destroyed. Such vandalisms reflected disharmony among farmers and system officials on the newly adopted mode of water distribution.

Canal desilting, which was relatively prevalent in Agos RIS, is the result of insufficiently taking into consideration in the original system design and subsequent rehabilitation or improvement works, the sediment load that would be expected especially for just a barrel-type main intake structure like that of the Agos RIS. However, provision of a silt excluder or any other structure to trap the silt and minimize its entry into the canal network was neither among the investment made in at least the past 20 years nor included in an on-going foreign-assisted project in Agos RIS.

The gabion dikes and concrete walls at the approaches of the headwork were constructed to protect the dams from strong floods. The construction of additional length of the crest of the diversion dam of Balanac RIS became an urgent task in 2010 for the IMO in-charge of the system. The damage caused by the flood spawned by a strong typhoon in that year had seriously compromised the stability of the dam. The construction of additional crest length to better anchor the dam on one side of the river bank was completed in 2013. This was the fourth time a crest segment of the dam was added.

The typhoon event underscored the need for a more reliable assessment of probable flood magnitudes for Balanac River for the purpose of planning and design of rehabilitation or improvement works. However, an analysis that would provide insights in the magnitudes of flood events associated with different chances of occurrence for the Balanac River has yet to be done. River gauging or discharge measurement has not been done since 1980.

In terms of investment made, access roads were given more emphasis than some key physical components of the irrigation systems. Roads had a much higher level of investment than the headwork and canal structures in Sta. Maria RIS. It also had higher investment than canal structures in Balanac RIS. Road projects have positive externalities for the local population. However, unlike the physical structures of irrigation systems, they have no direct contribution to the goal of expanding the actual area irrigated.

There was a significant difference in terms of rehabilitation expenses and magnitude among the three systems during the past 16 years (2000–2015). Balanac RIS had the highest rehabilitation expenses and highest reported total area rehabilitated while Sta. Maria RIS had the least for both.

6.6.3   Impacts of rehabilitation

The impacts of the rehabilitation and improvement projects carried out for the three NIS were assessed in terms of the trend in service area and actual irrigated area during the period 2000–2015 (Table 6.4). Agos RIS had the highest percentages of the design service area actually irrigated during both seasons, with average values of 80% and 81% for the wet season and dry season, respectively. For the Balanac RIS, the corresponding percentages for the wet and dry seasons were 70% and 76%, respectively. Only about 33% and 34% of the design service area of the Santa Maria RIS was irrigated during wet and dry seasons, respectively. The low percentage of irrigated area of Sta. Maria RIS was attributed by farmers and NIA staff to inadequacy of water supply from the dam to irrigate the whole design service area.

It is interesting to note that although the FUSA is significantly smaller than the design service area and adjusted periodically for design and operational constraints as manifested by operational reality, there was no season in any of the three systems that an entire FUSA has been served during the past 10 years. The three NIS more than 80% of their FUSA except during the 2015 dry and 2010 wet seasons for Balanac RIS and 2015 wet season for Sta. Maria RIS. The dam of Balanac RIS was heavily damaged by the typhoon and only 69% of its FUSA was irrigated during the 2010 dry season. The low percentages of FUSA actually irrigated in these systems in 2015 were attributed to the El Niño-induced drought.

Agos RIS had the highest percentages of actual area irrigated, averaging more than 90% for both cropping seasons, except in 2006. The main intake structure and upstream portion of the main canal of Agos RIS were heavily silted and rendered inoperable during the 2005 dry season by the record-high flash flood in late 2004.

Table 6.4. Firmed-up service area (FUSA) and irrigated areas (in 10^3 ha), by season

Year	Agos RIS[1]			Balanac RIS[2]			Sta. Maria RIS[3]		
	FUSA	Irrigated area		FUSA	Irrigated area		FUSA	Irrigated area	
		Wet	Dry		Wet	Dry		Wet	Dry
2000	1,280	1,280	1,280	1,056	850	970	-	-	-
2001	1,280	1,280	1,280	1,056	870	940	974	909	989
2002	1,280	1,200	1,280	1,056	847	946	974	844	844
2003	1,280	1,200	1,245	1,056	851	952	974	909	966
2004	1,280	1,232	1,232	1,000	850	950	974	876	876
2005	1,234	1,033	-	1,000	805	950	974	869	869
2006	1,234	1,033	1,033	1,000	850	970	974	837	837
2007	1,234	1,125	1,125	1,000	850	970	974	786	815
2008	1,234	1,150	1,150	1,000	870	940	974	837	837
2009	1,234	1,175	1,175	1,000	850	970	974	866	898
2010	1,234	1,190	1,100	1,000	850	690	974	856	858
2011	1,234	1,220	1,170	1,000	850	824	974	858	858
2012	1,272	1,258	1,258	1,000	850	850	974	866	866
2013	1,456	1,284	1,284	1,000	878	920	974	866	866
2014	1,456	1,284	1,284	1,000	796	920	974	842	857
2015	1,456	1,284	1,284	1,090	800	902	974	657	854

[1]Design area = 1,500 ha [2]Design area =1,200 ha [3]Design area =2,580 ha

There was no discernible change in the trend of actual area irrigated in Balanac RIS except in 2010 when the damaged dam resulted in a steep drop in the actual area irrigated during the 2010 dry season. The rehabilitation and improvement works during the period from 2000 to 2015 in the Balanac RIS covered at least 12,400 ha, which was around 12 times its FUSA. Yet, these considerable efforts had at best helped maintain an average actual irrigated area of around 870 ha during the 2006–2015 period or 86% of its FUSA. As of 2015, the dry season actual areas irrigated were way below their levels prior to 2010. Meanwhile, the percentages of FUSA actually irrigated during the wet season were declining in recent years.

A decreasing trend in the actual area irrigated can be observed in Agos RIS in both seasons during the period from 2000 to 2005. An abrupt drop in actual area irrigated from 2005 to 2006 was attributed to damage to the physical structures caused by a flash flood event in late 2004. The flash flood also resulted in heavy siltation of canals and service area and, consequently non-cultivation of many farms during the subsequent cropping seasons. A gradual recovery from the effects of the flash flood can be observed from 2006 onwards. A total of at least 5,700 ha (or about four times the service area) has been rehabilitated and restored and 30 ha has been generated during the period from 2006–2015. Starting 2013 the actual area irrigated was 1,284 ha, the highest value achieved in Agos RIS for the 2000-2015 period.

There was no discernible change in the actual area irrigated in the Sta. Maria RIS. With the rehabilitation and restoration efforts during 2000–2015 that covered about 5,300 ha, or at least five times of its FUSA, Sta. Maria RIS was able to irrigate an average of at least 827 ha or 85% of its FUSA during the 2006-20015 period.

The disparity between rehabilitation efforts and system performance among the three NIS is noteworthy. Balanac RIS had the highest total area rehabilitated and most rehabilitation investments. Yet, it had the lowest percentage of FUSA actually irrigated.

There are notable differences in the trends of data on actual area irrigated between that of the national level and the NIS studied. The percentages of the service area actually irrigated for both cropping seasons were significantly higher than the national average, which is about 73% and 71% during the wet season and dry season, respectively. Also, the actual area irrigated during the dry season for each of these three NIS was, in general, either higher or equal to its corresponding wet season value. In contrast, the wet season irrigated area of existing NIS in the country remained consistently higher than that of the dry season.

In general, it is expected that an irrigation system can irrigate a larger area during the wet season due to higher rainfall and river flows at the diversion sites. Some of the possible factors that resulted in the observed reverse trend for the three NIS could include the following: site-specific constraints (e.g., inundation during monsoon months); potentials to address limitations on surface water supply (presence of creeks and groundwater for augmenting water supply); availability of irrigation technologies (well drilling and development); and the capability and willingness of system managers and farmers to adopt an appropriate technology or course of actions.

In the sample NIS, some of the above mentioned factors have been circumvented to some extent. The farmers downstream in Agos RIS and Balanac RIS are using shallow tubewells to tap groundwater either for supplemental irrigation or as sole source of irrigation water during the dry season. Check gates were constructed for the purpose of tapping additional water from creeks and recapturing drainage water from upstream farms for reuse in downstream farms of Sta. Maria RIS and Balanac RIS.

The lower actual area irrigated during the wet season in the case of the Balanac RIS was attributed by NIA field personnel and farmers to the submergence of the portions of the service area bordering the Laguna Lake during rainy months. During the wet seasons of 2008 to 2011, flooded areas ranged from 46 to 300 ha of the service area. This inundation problem is yet to be addressed. Meanwhile, a strategy of starting the wet season cropping earlier than the traditional practice was adopted to circumvent a similar flooding problem in Sta. Maria RIS. Flood-prone rice areas are scheduled to be irrigated and planted first so that when the peak of the monsoon season comes the rice plants in these areas are already in their vegetative stage, thus, have better chance of withstanding intermittent floods or submergence. In Agos RIS excess water that enters the canal network drains into creeks through a spillway adjacent to the intake end of inverted siphons. Escape ways for surplus irrigation water were constructed in strategic points along the canal network.

6.6.4   *Potentials for increasing the irrigated area and for irrigation modernisation*

Agos River has a river basin area of at least 879 km^2 and an estimated minimum dry season surplus of 67.8 MCM (million cubic metres) and 19.1 MCM at the start of February and at the end of May, respectively (NIA, 2006). Thus, it is reasonable to assume that a higher percentage of actual area irrigated in Agos RIS is not constrained by the inadequacy of water supply at the source. This view was consistent with the view of system officials and farmers that Agos River has enough flow to meet the irrigation requirement of the entire service area even during the dry season. However, unlike Balanac RIS and Sta. Maria RIS that both have a diversion type dam, Agos RIS has only a 2 x 2 m^2 barrel opening at the river bank as main intake structure. A temporary brush dam is built at the start of each dry season to increase river flow into the main intake structure. The high river flows during the rainy season wash away such a dam, hence the need to construct a new one. The construction of such a brush dam has become the regular major activity of the NIA field office. Such diversion method becomes the limiting factor in diverting water into the service area during the dry cropping season. Also, it did not have provision to prevent sediment transport into the main intake structure.

A number of downstream canals of Agos RIS were already inoperable due to heavy silt deposit, which was exacerbated by the record-high flash food in 2004. The silt that accumulated in the canals had practically buried the canals including the flow control structures. Re-canalization would be needed to improve irrigation to downstream service area. Moreover, a mechanism to minimize entry of sediment through the main intake structure or trap it before it is carried farther into the canal network would need to be put in place.

By design, water supply of Sta. Maria RIS was limited. The design diversion capacity of 0.8 l s^{-1} ha^{-1} for Sta. Maria RIS is just about half of the 1.5 l s^{-1} ha^{-1} water requirement NIA adopted in its design guidelines for lowland rice irrigation. This inherent low water supply could be the main reason behind the 1,600 ha difference between the design service area and the FUSA. Even though the service area was reduced from the original 2,580 ha to 974 ha, system officials have not been able to irrigate the entire FUSA. They attributed this shortfall to the low dry season flows of Sta. Maria River to meet the irrigation requirement of the service area. They are addressing their water scarcity problem by lining the canals to increase conveyance efficiency, reusing drainage water from upstream farms, staggering planting dates within the service area and by installing water distribution structures to enforce the agreed irrigation schedule for the different turnout service areas.

Preliminary results of a complementary study on the water balance of Sta. Maria RIS suggest a relatively efficient use of diverted water. About 108 ha of the service area is irrigated using drainage water from upstream farms within the service area and nearby farms outside the service area. Meanwhile, springs and minor waterways from nearby mountains are tapped for irrigation water for about 78 ha of the service area. It is evident that the potential for increasing the actual area irrigated through a more efficient use and management of diverted water is quite limited. A substantial increase in irrigated area can be realized with an increase in water supply.

A number of springs can be found along the foot of the mountain surrounding the service area of Sta. Maria RIS. Development of small irrigation systems that will tap spring water presents an opportunity to augment the water supply. It would be worthwhile to study the feasibility of spring development for irrigation for future improvement projects.

The quantity of water diverted for Balanac RIS is not being measured. Based on the system profile document and field observation, there would be sufficient supply of water into the Balanac RIS to meet its irrigation requirement. This view is supported by a common response of farmers and system officials to questions relating to adequacy of river flows and diverted water based on their own observation and experience with the dam and canal operation.

There are compelling reasons to assume that the cause of inability to irrigate the whole FUSA of Balanac RIS during the dry season lies in water management and distribution. System operation carried out by the NIA field staff in Balanac RIS mainly involves opening and closing of gates of the main intake structure at the dam. Basically, the system operation is for continuous irrigation. Farm ditches directly offtaking from the main through ungated turnouts are prevalent. Also, many head control structures at major distribution points along the main canal are either in disrepair or missing the spindle and metal plate. Most offtakes of secondary canals and major turnouts are ungated. The use of makeshift gates made of stop logs, bamboo poles and banana trunks is common in the area. Consequently, irrigation water would be available to downstream farmers whenever it is not needed upstream. However, this setup does not work well with farmers downstream especially during the dry season. Downstream farmers faulted their upstream counterparts for diverting water much more than what is needed and for wasteful use of diverted water.

The potentials for achieving full service area coverage in Balanac RIS are in improvement of management and delivery of irrigation water. Flow control structures necessary to achieve a more efficient water management would need to be installed to help curtail any unreasonable rates of water diversion in upstream farms, thus, increase water available to downstream farms during the dry season. A rule-based water allocation scheme aimed at irrigating the whole service area would need to be formulated.

Balanac RIS had the lowest percentages of FUSA irrigated during wet seasons among the three systems studied. It also had the biggest difference in actual area irrigated between the wet and dry cropping seasons. This difference was attributed to the inundation of the service area adjacent to the Laguna Lake. Pilot testing strategy adopted by system officials and farmers of Sta. Maria RIS for a similar inundation problem would be a sensible course of action for Balanac RIS. A shift in cropping calendar for the downstream service area to avoid flood months or minimize the detrimental effects of floods on crops would make cultivation of normally submerged farms possible.

The lack of data on river discharge at diversion points, diverted water for irrigation, rainfall and other climatic parameters for estimation of evapotranspiration, seepage and percolation rates poses a challenge to a water balance study whose results could indicate the potentials for irrigated area expansion or irrigation system modernisation. Flow or head measurements at major water control or distribution points along the canal network and parcellation maps of different irrigation zones that would be useful in identifying critical points and bottlenecks in achieving water delivery targets are non-existent. Separate studies would need to be carried out first to generate data needed as inputs in the formulation of modernisation plans.

6.7 Conclusion

The continued mediocre performance of canal irrigation systems has been a cause for an action in a country striving to achieve rice self-sufficiency and food security. This performance and the escalating cost of canal irrigation development and the deterioration of existing systems were the primary reasons for the mandated public expenditure shift towards rehabilitation and improvement of existing irrigation systems and the development of effective, affordable and efficient minor irrigation systems by the Agriculture and Fisheries Modernisation Act.

Available information on rehabilitated, restored and actually irrigated areas suggests that there are problems in the planning, design and operation and maintenance of canal irrigation systems, which are largely unaddressed during rehabilitation and improvement. Even with available water supply and considerable rehabilitation efforts, the average percentage of O&M service area actually irrigated in both seasons during the past 10 years averaged 73%. It is compelling to examine the root cause of these problems.

The inclusion of water adequacy among the criteria for prioritization of rehabilitation projects for funding approval reflects recognition to the fact that the availability of water for the proposed rehabilitation, restoration and expansion of service area is important. Indeed, availability of water is a precondition for realizing the benefits of rehabilitation or improvement works. Therefore, it needs to be ascertained before a proposal can be considered in the prioritization process. The merit of the proposal in terms of the water adequacy criterion would have to be supported with a water assessment study or water flow data.

A strategic monitoring and evaluation system to ensure quality of construction and to assess the performance of rehabilitation projects against their stated targets would have to be put in place to promote accountability among the parties involved in the project planning, design and implementation and to serve as a feedback mechanisms for policy- and decision makers. Farmers and independent research groups or institutions would be viable partners in a monitoring and evaluation process.

Agos RIS, Balanac RIS and Sta. Maria RIS had consistently higher percentages of FUSA irrigated than the national average. Such a performance could be attributed to the improvement works that have addressed the apparent design shortcomings in their respective systems and also to the measures adopted to circumvent the inherent and site-specific constraints. Some of the notable improvement efforts included the use of tubewells for irrigation, provision of canal structures for drainage water reuse and water distribution and a shift in cropping calendar and provision of escape way for excess water to adapt to prevailing weather patterns. The experiences in rehabilitation and improvement in these three irrigation systems demonstrated what would work and provided insights of what would have to be done to increase the chance of closing the gap between the service area and the actual irrigated area. The case of Sta. Maria RIS showcased the utility of flow control structures and a shift in cropping calendar in overcoming its water constraints.

A more systematic approach to rehabilitation of irrigation systems would need to be pursued. There is a need to develop a modernisation plan for each irrigation system. Site-specific potentials and constraints to irrigation modernisation need to be taken into account in the formulation of such a plan. Diagnostic assessment of the physical structure and operation and maintenance of irrigation systems and field validation of design values of water balance parameters need to be carried out to identify constraints and prioritize improvement options.

7 DIAGNOSTIC ASESSMENT OF THE IRRIGATION SYSTEM DESIGN AND PERFORMANCE

7.1 Introduction

The continued performance of national irrigation systems below the expected levels despite rehabilitation and improvement efforts have been a subject of discussions by concerned government policy- and decision-makers. Many questioned what constrains the performance of irrigation systems and what intervention measures should be implemented. There have been assertions coming from different stakeholders on the reasons of poor performance of irrigation systems. Irrigation professionals, academicians and individual experts, system operators and farmers groups have a number of contrary opinions on what are the problems and effective solutions.

While everyone would easily agree that each of the problems raised contribute to mediocre performance, a consensus on what is the right approach to address these constraints is hardly reached. This may be due to the fact that many stakeholders are coming from different viewpoints and have different expectations. Many may have first-hand experience with specific constraints or causes of a problem at a particular tier of planning, management and operation of irrigation systems. However, clear-cut common understanding of the interrelations of individual constraints and comparative contribution to the problem of poor irrigation system performance as a whole is lacking. There is a need for a holistic view on the irrigation systems to clarify the causal relationships among the observed problems. Analysis of the mechanisms and processes of water delivery at the system-level would distinguish the root causes from intermediate effects. Only after the root causes are identified an appropriate approach can be formulated.

Diagnostic assessment is a performance evaluation method specifically aimed at identifying the bottlenecks of irrigation system performance and, hence, an appropriate approach and solutions to address them. It has been carried out (Burt and Styles, 1999; Kumar et al., 2010; Garcia-Bolanos et al., 2010; Bruscoli et al., 2011). It is considered a crucial prelude in the modernisation process (Renault et al., 2007).

The planning and formulation process for system improvement projects in the Philippines does not include a diagnostic assessment (Delos Reyes *et al.*, 2015). Hence, diagnostic evaluation is rarely carried out for government-funded canal irrigation systems. This lack of emphasis on the importance of a sound diagnosis of irrigation systems is unfortunate, especially when the development of most canal irrigation systems represent a typical case that would greatly benefit from a diagnostic study. Most NIS have undergone a mix of technical, management, policy and institutional changes whose interplay and effects could have provided leads on how the system performance can be improved had they been well understood. During different periods since the early 1990s, irrigation development of the country was aimed at achieving a combination of broad societal goals such as food security, rice self-sufficiency, increased farming productivity and income, creation of livelihood opportunities and acceleration of development in rural areas. The planning and design of large-scale NIS projects involved foreign expertise and design guidelines and irrigation technologies based mainly on the United States Bureau of Reclamation standards. Implicit to the design of the physical structures of these systems were operational and institutional requirements, which proved to be beyond what the Government can sustain (World Bank, 1982, 2006; David, 2003; Asian Development Bank, 2007, 2011). The NIA modified and replaced the physical components of most irrigation systems as necessitated by clamours for improved water delivery, changes in water demand and allocation, shift in irrigation development thrusts and policy directions of the Government, as well as, the requirements of international financing institutions. Recent system improvement and modernisation projects have replaced the gated, adjustable flow control structures with fixed weirs. Also, the NIA promotes the IMT, through which, many of the water distribution responsibilities have been transferred to farmer's groups. The IMT program aims for a gradual transfer of full system management to IAs. With such mix of changes in institutional and technical aspects of irrigation systems, the more compelling it is to conduct system diagnosis to examine the interactions or interdependence among different system components, irrigation water requirements and system operations and to better understand the implications of the changes in any of them. Consequently, the root causes can be systematically identified. Such findings of diagnostic assessment are relevant inputs in the formulation of a modernisation plan and design of system improvement works.

This chapter explores the utility of coupled diagnostic assessment techniques deemed relevant in the formulation of a modernisation strategy for many NIS in the Philippines. This diagnostic technique is a combination of a design logic framework advocated by Ankum (2001) and diagnostic tools used in the mapping system and services for canal operation techniques (MASSCOTE) (Renault *et al.*, 2007) in diagnosing the performance of small-scale NIS in the Philippines. In the design logic framework, the coherence of the philosophy, objectives of system operation and design of physical structures are examined. The physical and operational constraints of performance that are inherent to the system design are identified.

The design logic framework of analysis recognizes the fact that irrigation systems were built to realize a set of irrigation objectives in particular socio-political and agro-climatic settings. The water supply availability and existing water laws, water rights or water distribution philosophy of the society for meeting irrigation need of the population shape the design framework of irrigation systems. For a stated system objective, certain design parameters, system configuration and combinations of flow control structures are more logical than the others. For example, a maximum crop production would need reliable water supply. It would necessitate a storage-type dam in cases where the water sources have significant seasonal fluctuations. The a priori rights or first-in-use rights consideration would require some flow control structures while riparian rights might not require any. Similarly, a proportional sharing of water supply would mean less physical components to regulate the flow.

Also, the design logic framework tackles examination of the design configuration and canal structures, which largely determine the options for operations and influences the hydraulic behaviour of the canal system. In gated irrigation systems, the hydraulic characteristics and operation of the selected combination of flow control structures can either facilitate or hamper achieving the water delivery targets. The net effect depends on how the canal structures behave under unsteady state flow conditions and perturbations or significant changes in canal discharges and water levels occurring along a canal network as a result of intended or unpredicted changes in inflows or outflows, adjustments in the setting of structures, or transient flow during distribution changes. Different types of water level control structures exhibit different levels of sensitivity or change in observed head for a given change in incoming discharge. Same case is true in the sensitivities of canal offtakes and turnouts. The more sensitive structures generate or amplify perturbations or a significant change in flow along a canal network. They require more frequent resetting or adjustments than less sensitive structures to keep water delivery within an acceptable range of the delivery targets. Ungated systems and those under the principle of proportional flow distribution do not have such concerns.

The diagnostic tools of MASSCOTE, which include the rapid appraisal procedure (RAP) and assessments of physical capacity and hydraulic behaviour (sensitivity and perturbations) of irrigation systems, are aimed at systematic and comprehensive evaluation of the performance of large-scale canal irrigation systems to modernize canal operation (Renault *et al.*, 2007). They consider the water balance, internal processes and mechanisms (hardware, operational procedures, management and institutions) of water delivery at different canal levels, irrigated area and crop production in diagnosing the system performance. The actual performance of the system, as built and operated is gauged from the perspective of service-oriented management.

By the nature of their respective analytical approaches, it is deemed that a combination of the two techniques would lead to better segregation of the constraints of performance with their respective root causes – whether they are due to design, operation or unrealistic expectations. The coupled diagnostic performance assessment was carried out for Balanac RIS and Sta. Maria RIS. The findings of the assessments are discussed in this chapter. Identification of the root causes of performance will help in the formulation of appropriate intervention measures for consideration in the formulation of modernisation strategies for these systems and similar irrigation systems in the country.

7.2 Assessment methodology

The first part of the diagnostic assessment critically analyzed the logical coherence of the stated irrigation system objectives, design philosophy, flow control structures, system operations, water supply and water delivery targets of Balanac and Sta. Maria RIS based on the analytical framework (Figure 4.4) proposed by Ankum (2001). The consistency among them was evaluated based on established logical combinations of flow control methods (e.g.: upstream control, proportional control), operational objectives of the main system (e.g.: method of water allocation to offtake service area and method of water distribution through main canals) and the stated design philosophy (e.g.: flexible supply based on water need and availability, proportional allocation of water supply based on farm size). The appropriateness of the flow control structures was evaluated based on recognized logical combinations of hydraulic structures to meet the stated operational objectives (imposed or semi-demand allocation, splitted, intermittent or adjustable flow to offtake service area; splitted, rotational flow or adjustable flow through the main canals) and on the ease or simplicity of required canal operations. The project feasibility study and original project design documents, which usually contain the required information to characterize the case study systems according to the design logic framework, were sought. The information on these documents provided a perspective on what was originally intended, how well the means can achieve the goal, and how coherent the succeeding modifications or improvement works are to the original or the present objectives. Unfortunately, copies of the project feasibility and project design documents for Balanac RIS and Sta. Maria RIS were no longer available at the Central Office and field offices of the NIA. Hence, such information was gleaned from available, more recent project documents for the system under study, design manuals or guidelines of the NIA, system walkthrough, and informal interviews with design and operation staff of NIA and water tenders of irrigators associations.

The second part of the diagnostic assessment involved field investigation of the current state of the physical structure and canal operation of the selected NIS. This included the conduct of the first three steps in MASSCOTE: the Rapid Appraisal Procedure (RAP), system capacity and sensitivity assessments and perturbation analysis. The required data input for the calculation of RAP indicators of system performance were gathered from system documents and the weather agency. They were also obtained through interviews of irrigation managers, operation staff, irrigators associations and system walkthroughs. The interview questions related to their perceptions on how the irrigation system functions – what levels of irrigation service the main canal delivers, what the operators do and how water reaches individual farmers. A walk through the canal network, from the dam to the end checks of main canals and second-level canals or major bifurcation canals were carried out for each NIS. The actual methods of operation, level of maintenance and general condition of the hardware or physical structures (headwork, canals and flow control structures) were observed during the walkthroughs. The system capacity was assessed in terms of the physical capacities of the irrigation system to perform its functions of conveyance, diversion, division, water level regulation, flow measurement, storage and transfer capacity. The actual capacity for each function was compared to the design capacity and the current required capacity. Also, the origins of perturbations were identified for the purpose of determining whether the main canal is self-reacting or if specific interventions (e.g.: adjustments in system operations and management or changes in flow control structures) must be carried out to improve the water distribution.

7.3 Assessment of the consistency among the design philosophy, operational objectives and physical structure of the irrigation system

A review on the general guidelines and practices on irrigation system planning and design was carried out to gain insights on the design philosophies and goals adopted in the country. The findings include the following design considerations and objectives:

- *rice monoculture*. Self-sufficiency in rice, the staple food in the country, has been the policy of the Government. Canal irrigation systems are designed to supply water for paddy rice cultivation during the wet and dry cropping seasons;

- *productive irrigation*. The crop water requirement during periods of favourable climatic conditions to rice production was the foremost consideration in determining the capacities of canals and flow control structures. This implies that the irrigation systems were intended for productive irrigation, that is, to provide water for optimum crop growth;

- *dry season as the design main irrigation season*. The highest irrigation water requirement for a unit duration of the cropping pattern is considered as the design irrigation water requirement. As the country receives abundant rainfall during the wet season, the highest irrigation water requirement is usually obtained during the dry cropping months. This design approach suggests that the irrigation system is aimed at satisfying the crop water requirement in the service area during the dry season;

- *equitable supply and flexible supply.* The use of water duty (flow rate per ha) as a design criterion suggests a design philosophy of equitable supply of irrigation water to a unit 'area. Canals and flow control structures are designed to operate under two flow regimes: high-flow condition during land soaking and preparation; and low-flow condition during the crop maintenance period. Adjustable gates are selected to effect this amenability for flexible water delivery;

- *imposed allocation and upstream control.* The final water distribution plan and operational procedures are decided by system officials. The NIA system operators are responsible for managing the dam and main intake operation and have the authority on the control of head gates and other flow control structures along the main canal and other major conveyance canals, which include the second- and third-level canals in medium- to large-scale systems. Cross-regulators are used to maintain the upstream water level at the levels that correspond to the target offtake or turnout discharges. Flumes and weirs were constructed for discharge measurement.

7.3.1   The original design of Balanac RIS and Sta. Maria RIS

The Balanac RIS and Sta. Maria RIS schemes were designed to irrigate paddy or lowland rice in their respective municipalities. Their diversion dams are of run-of-the-river type with at most two manually operated vertical steel gates for the culvert-type main intake structure and sluice gates for flushing out sediments (Figures 7.1-7.3). Balanac RIS has a bifurcating canal layout (Figure 7.4) while Sta. Maria RIS has a hierarchical one (Figure 7.5). They were designed as gated system as indicated by gated flow control structures along their respective canal networks and flow measurement structures upstream of secondary canals and at some points along the main canal.

A combination of concrete pipe farm turnout, concrete pipe bifurcating canal intake and cross regulator or water level control were originally constructed at major bifurcation points in Balanac RIS (Figure 7.6a). Each of these flow control structures were with spindle-type vertical steel gate and a pair of premade guides or grooves on the side wall of the concrete upstream approach.

(a)   (b)

Figure 7.1. The sluice gates (a) and main intake structures (b) of the Balanac Dam

(a)   (b)

Figure 7.2. The Bagumbayan Dam (a) and its main intake structures (b) of Sta. Maria RIS

(a)   (b)

Figure 7.3. The Mata Dam (a) and its main intake structure (b) of Sta. Maria RIS

In Sta. Maria RIS, a constant head orifice (CHO) as farm turnout, a concrete pipe culvert with vertical circular gate as secondary canal intake and a vertical steel gate as cross regulator were installed at its major flow distribution points especially along its main canals (Figure 7.6b) All of its vertical gates were of spindle type. Its secondary canal intakes had a similar construction to that of a CHO, except that the upstream gate had premade guides similar to those slots for stoplogs instead of a submerged orifice.

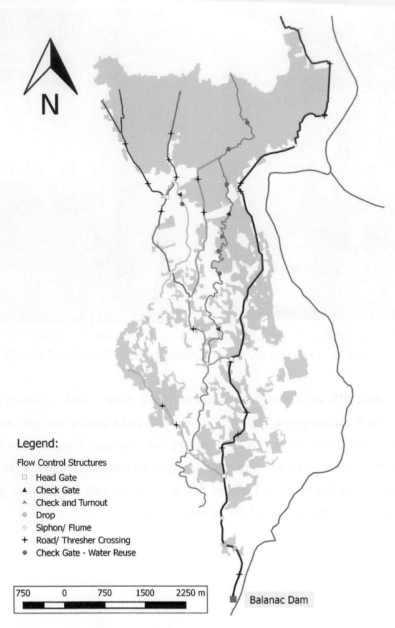

Legend:

Flow Control Structures
- ☐ Head Gate
- ▲ Check Gate
- ⌃ Check and Turnout
- ⊙ Drop
- ◇ Siphon/ Flume
- + Road/ Thresher Crossing
- ● Check Gate - Water Reuse

750   0   750   1500   2250 m

Balanac Dam

Figure 7.4. General layout of Balanac RIS canal network with flow control structures

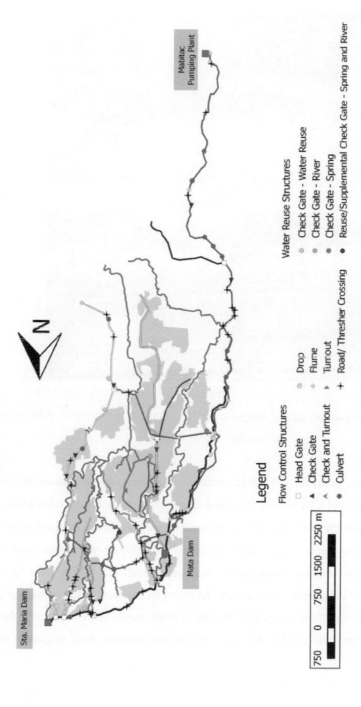

Figure 7.5. General layout of Sta. Maria RIS canal network with flow control structures

(a)

(b)

Figure 7.6. Cross regulator, offtake and farm turnout along the main canals of Balanac RIS

(a) and Sta. Maria RIS (b)

In both systems, gated circular culverts without immediate cross regulators (that is, free offtake arrangement) and CHOs were also used as minor turnouts and offtakes along canal reaches between major distribution points of the main canals and second-level canals or major bifurcating canals. Also, the cross regulators at major distribution points of both systems were combined with a drop structure making canal reaches hydraulically independent from each other. The offtakes of dependent canals at these junctions were either set lower than the parent canal bottom or with significant drop also. In both systems, the design cross section for main distributary canals was trapezoidal. All canals of the two systems were originally earthen canals.

7.3.2   The present design of Balanac RIS and Sta. Maria RIS

Rehabilitation and installation of additional structures have been carried out in the two systems since their first years of operation (Tables 7.1 and 7.2). In the late 1970s, pumping stations and the corresponding canal network were constructed to irrigate rice areas within the vicinity of the present Sta Maria RIS. This pump irrigation scheme had long been abandoned mainly because of the prohibitive costs of pumping. A more recent addition to the dam of Sta. Maria RIS was the gabion wall for upstream protection of its right abutment. The major changes in the headwork of Balanac RIS were the added lengths or segments of the crests of the dam and the concrete protection wall of the dam approach (Figure 7.7). The manual lifting mechanisms for the sluice gates of the dams in the two systems were changed to mechanical.

In the 1990s, a system modality improvement project in Balanac RIS replaced the spindle-gated offtakes and water level regulators at major bifurcation points with duckbill and long-crested, proportional weirs. It also embarked on filling up portions of the cross sectional areas of ungated culvert-type farm turnouts along the main canal with concrete cement to reduce their intake capacities. The reduction in size of ungated offtake openings was based on their respective irrigation water requirements or size of service areas. Only few remnants of the spindle-type vertical gates of the offtakes, farm turnouts and cross regulators, usually the steel frames and the accompanying concrete structures were left at major distribution points along the main canal and branching points of lower level canals.

In contrast, Sta. Maria remained a gated system. Most of the original flow control structures at major diversion points along its main canals are still in-use. Additional vertical steel gates to raise the water level were installed at the downstream portion of Sta. Maria RIS main canals to facilitate water distribution. Also added to the original structure were minor offtakes with non-modular stopboards as gates. They were situated along main canals and second-level canals. Its service area was reduced to 40% of the original.

Table 7.1. System characteristics pertinent to system design and operational objectives of Balanac RIS

System characteristics[1]	Original design	Design changes
Approved diversion rate, 1 s[-1]	3,900	--
Design service area, ha	1,200	1,000
Source of water supply	River	--
Water supply at the source		
Wet season	Abundant	--
Dry season	Sufficient	--
Diversion dam	Run-of-the river (ROR) dam	Lengthened dam crest, protection walls
Dam intake gates operation	Vertical, manual	--
Dam sluice gates operation	Manual	Mechanical
System water use	Rice irrigation	+ few orchards
Share of irrigation supply	Equitable supply to area	Equitable supply to area
Gate setting at the tertiary offtakes	Possible at fixed time	Not possible
Water allocation to tertiary unit	Imposed	--
Controlled variable	Discharge	--
Offtakes/turnouts along main System	Orifice with spindle-type vertical gates	Ungated orifice, weirs
Measurement structures	Sharp-crested rectangular weirs	--
Water level regulators along main system	Spindle-type vertical gates	Duckbill, long-crested, proportional weirs
Freedom and precision of control of structures at the main system	Gradual	Fixed Stepwise
Flow control at main system	Upstream control	Proportional control
Access/manipulation of structures	Manual, in-situ	Manual, in-situ
Immediate storage	None	--
Drainage system	None	Low
Water reuse check gates	None	7
Conjunctive use	None	+ pump irrigation
Canal materials	Unlined	+ lined

[1]Modified typology based on Renault and Godaliyadda (1999); -- (none) + (added)

Table 7.2. System characteristics pertinent to system design and operational objectives of Sta. Maria RIS

System characteristics[1]	Original design	Design changes
Approved diversion rate, l s^{-1}	2,100	--
Design service area, ha	2,500	974
Source of water supply	River	+ creek, spring
Water supply at the source		
Wet season	Limited	Limited - short
Dry season	Limited	Short
Diversion dam	ROR dam	+ pump station
		+ protection walls
Dam intake gates operation	Vertical, manual	--
Dam sluice gates operation	Manual	Mechanical
System water use	Rice irrigation	+ few orchards, fish
		culture and livestock
Share of irrigation supply	Equitable supply to area	--
Gate setting at the tertiary offtakes	Possible at fixed time	--
Water allocation to tertiary unit	Imposed	--
Controlled variable	Discharge	--
Offtakes/turnouts along main	Orifice with spindle- type	+ ungated orifices,
System	vertical gates	weirs
Measurement structures	Parshall, cutthroat flumes	--
Water level regulators along main system	Spindle type vertical	--
	gates	
Freedom and precision of control	Gradual	+ On/off,
of structures at the main system		Stepwise, Fixed
Flow control at main system	Upstream control	--
Access/manipulation of structures	Manual, in-situ	Manual, in-situ
Immediate storage	None	--
Drainage system	None	Low
Water reuse/augmentation check gates	None	9
Conjunctive use	None	+ River, spring
Canal materials	unlined	+ lined

[1]Modified typology based on Renault and Godaliyadda (1999) -- (none) + (added)

(a)

(b)

Figure 7.7. The lengthened crest of the Balanac Dam (a) and the gabion protection wall of Bagumbayan Dam in Sta. Maria RIS

Seven and nine water reuse structures (check gates) were installed in Balanac RIS and Sta. Maria RIS, respectively, to capture drainage water for reuse in downstream farms. Also, many ungated culverts and open, rectangular offtakes on the canal walls were constructed along with concrete canal lining works. The former were of an orifice type, while the latter were either of weir type (with crest or raised sill) or conveyance type (same canal bottom elevation). About 60% and 75% of the total canal lengths of Balanac RIS and Sta. Maria RIS, respectively, were lined with concrete. The details on some relevant physical features of the two systems are summarized in Tables 7.3 and 7.4. They were obtained from various Laguna-Rizal IMO files, WUAs records and through system walkthrough and on-site measurements of the structures.

Table 7.3. Pertinent physical features of Balanac RIS

	BMC	Buboy	Lat. A	Lat. A1	Lat. A1A	Reuse
Headgate service area, ha	812.8	103.7	323.1	260.6	172.3	202.2
Total TO service area, ha	385.9	103.7	62.5	88.4	172.3	202.2
Discharge m^3 s^{-1}, U	1.89	0.08	0.96	0.57	0.15	Ni
Discharge m^3 s^{-1}, L	0.94	0.01	0.15	0.33	0.15	Ni
Flow area m^2, U	2.78	0.29	1.47	0.92	0.5	9
Flow area m^2, L	1.47	0.07	0.33	0.5	0.5	4
Canal length, km	14.9	2.8	4.9	4.8	2.7	5
Lined canal, %	87	66	92	76	94	0
Total TO	118	22	39	19	23	ni/nr
Official TO	45	20	20	10	16	8
Spindle-gated TO	0	0	0	0	0	0
Gated cross regulator	0	0	0	0	0	0
Main farm ditch	18	0	12	5	4	0
Measurement structure*	0	0	0	0	0	0
Water reuse check gate	0	0	0	0	0	7
Spill point	0	0	0	0	0	0
Drainage	PIN	PIN	PIN	PIN	PIN	PIN
Service road, km	6.0	0	0	0	0	0
Sediment control	0	0	0	0	0	0
Road/utility crossing	12	3	5	4	5	0
End check	1	1	1	1	1	0

Source: various Laguna-Rizal IMO files, WUAs records and system walkthrough; italics values are based on field measurement of physical structures.

*In-use; BMC - Balanac Main Canal; Lat. - Lateral; U - upper value; L - lower value; TO - turnout; PIN - paddy-to-paddy, irrigation canal, natural waterways; ni - no information; nr - not regulated

Table 7.4. Pertinent physical features of Sta. Maria RIS

	SMMC	Lat. A	Lat. B	Lat. C	Lat. D	MMC	Lat. E	Lat. F	Reuse
Headgate service area	459.1	246.6	72.8	32.5	18.2	513.7	141.5	69.0	201.3
Total TO service area h	89.0	246.6	72.8	32.5	18.2	303.3	141.5	69.0	201.3
Discharge U	2.20	0.58	0.53	0.37	ni	1.09	0.34	1.14	ni
Discharge L	0.79	ni	ni	ni	ni	0.84	ni	ni	ni
Flow area U	2.77	*2.09*	*0.53*	*1.43*	*1.37*	2.87	*1.64*	*0.53*	*18.35*
Flow area L	1.41	*0.62*	*0.15*	*0.39*	*0.48*	0.46	*0.57*	*0.53*	*2.75*
Canal length	6.6	7.5	1.8	2.3	0.6	7.2	2.4	1.9	5.49
Lined canal, %	53	100	100	28	96	85	54	100	61
Total TO	41	77	10	15	4	82	15	29	ni/nr
Official TO	21	51	6	14	4	35	7	0	11
Spindle-gated TO	11	8	1	0	0	14	0	0	2
Gated cross regulator	4	9	1	1	0	11	1	0	0
Main farm ditch	17	8	1	1	3	8	2	2	0
Measurement structure*	1	0	0	0	0	1	0	0	0
Water reuse check gate	0	0	0	0	0	0	0	0	9
Spill point	1	0	0	0	0	0	0	0	0
Drainage	PIN	PIN	PIN	PIN	PIN	PIN	PIN	PIN	PIN
Service road	6.6	7.5	1.8	2.3	0.6	7.2	0	1.8	0
Sediment control	0	0	0	0	0	0	0	0	0
Road/utility crossing	14	9	3	2	2	8	0	0	0
End check	0	0	0	0	0	0	0	0	0

Source: various Laguna-Rizal IMO files, WUAs records and system walkthrough; italics values are based on field measurement of physical structures.

Units: service area (ha); discharge (m^3s^{-1}); flow area (m^2); length, road (km)

*In-use; SMMC - Sta. Maria Main Canal;   MMC - Mata Main Canal; Lat. - Lateral; U - upper value; L - lower value; TO - turnout; PIN - paddy-to-paddy, irrigation canal, natural waterways; ni - no information; nr - not regulated

7.3.3   The present state of the physical structures of Balanac RIS and Sta. Maria RIS

This section focuses on the discussion on the systems components whose current state deviates significantly from their design or newly-constructed state and, hence, are no longer functioning as intended. The information presented was obtained from the findings of the system walkthroughs carried out for each of the systems as part of RAP.

The lifting mechanisms to operate the mechanical sluice gates used for water level control and flushing out of sediments at the main intake structures of Balanac Dam and Sta. Maria dams were not operational. The lifting gears have remained untouched in gear houses and have gathered rusts. The framed face and gate slides of the sluice gates were deformed and misaligned. The sluice gates of Balanac Dam are manually lifted by using a cable while those of Sta. Maria dams remained closed and had not been used since they became non-functional. The manually operated vertical gates of the main intake structures of Balanac Dam and Sta. Maria dams were still functioning, albeit they could not be fully closed due to a bend either in the framed face or spindle.

The fixed, long-crested and proportional weirs constructed to replace the adjustable gates of offtakes and cross regulators in Balanac RIS were all vandalized with some weir lengths chipped off (Figure 7.8). Many unauthorized ungated offtakes were found along the canal network. Significant lengths of the canals had broken or cracked concrete and grouted riprap linings. The flow measurement structures were no longer used. The concrete sharp crested weir immediately downstream of the main intake structure was submerged under normal flow conditions.

Meanwhile, most of the spindle-type, vertical gates of offtakes and cross regulators along the main and secondary canals of Sta. Maria RIS remained functional (Figure 7.9). The CHO at main distribution points were missing the adjustable steel component of their upstream gates; only the concrete structure forming the rectangular orifice remained. The downstream gates were still operational. The CHO along stretches of main canals either lost one or both of their gates while most of the circular vertical gates remained functional. There were also unofficial ungated offtakes along distribution canals. The concrete canals have cracks though not as rampant as in Balanac RIS. The Parshall flumes immediately downstream of the main intake structures of the Bagumbayan Dam and Mata Dam were overlain with considerable volume of sediments either at their respective approach canals or bottom of the throat sections. The flumes constructed immediately downstream of secondary canal offtakes were no longer used. All flumes, except for one, were missing the staff gauge.

Figure 7.8. Damaged long-crested weir and unathorized turnout in Balanac RIS

7.3.4 The consistency of physical structure and present water management practices with systems objectives and operational objectives

The primary objective of Balanac RIS and Sta. Maria RIS was to provide water to rice monoculture. This still holds true in both systems. The result of categorization of the system profile data based on the design framework of Ankum (2001) suggested that the original Balanac RIS was designed for productive irrigation for the dry season with flexible supply and imposed allocation to tertiary units or irrigation service delivery points (Figure 7.10).

The design water duty of Balanac RIS was $3.25 \, l \, s^{-1} \, ha^{-1}$ and was more than sufficient to meet the water requirements for lowland or paddy rice cultivation in the Philippines during dry seasons. The adjustable and stepwise gates for offtakes and cross regulators at major distribution points as parts of the original physical structure indicated that the design operational objective of Balanac RIS was to provide flexible irrigation supply to tertiary areas with adjustable flow through the main conveyance system. The use of cross regulators implied upstream control as the intended method for regulating water flow levels. Imposed allocation to tertiary units, or offtakes downstream of which the water management is no longer a responsibility of the NIA, has been the water allocation practice in the country. From the perspective of design logic framework (Figure 4.4) by Ankum (2001), there was coherence among the design philosophy, overall system objectives, operational objectives and flow control method in the original system design of Balanac RIS.

Figure 7.9. Cross regulator, CHO and offtakes along Sta. Maria RIS main canal

The system objectives for the original Sta. Maria RIS were the same as those of Balanac RIS, except in the case of the design irrigation season (Figure 7.11). With the design diversion capacity of 2.1 m^3 s^{-1} and service area of 2,500 ha, its water duty was 0.8 l s^{-1} ha^{-1}. This water supply was much lower than the 1.5 l s^{-1} ha^{-1} typically used by NIA for paddy rice cultivation.

Parameters	Choices for design philosophy/system objectives				
Irrigation target	Protective			**Productive**	
Cropping system	**Rice mono cropping**			Diversified cropping	
Design irrigation season	Wet season			**Dry season**	
Irrigation during other season	Dry season: all area - fallow small area - protective	Dry season: all area - protective small area - productive		**Wet season: Supplementary**	
Irrigation supply in the system	Equitable supply to farmers	*Equitable supply to hectares*		**Flexible supply based on water need and availability**	Flexible supply based on water demand
	Choices for operational objectives of main system				
Decision-making on water allocation to tertiary unit (TA)	**Imposed allocation**			Semi-demand allocation	On-demand Allocation
Method of water allocation to TA	*Splitted Flow*	Intermittent flow	**Adjustable flow**	Adjustable or intermittent flow	Adjustable or intermittent flow
Method of water distribution through the main system	*Splitted Flow*	Rotational flow	**Adjustable flow**	Adjustable flow	Adjustable flow
Flow control method	*Proportio-nal control*	**Upstream control** or simultaneous control		Upstream control or predictive control	Downstream or volume control

Logical design choices for: protective irrigation (gray shade); productive irrigation (green shade); possible for either protective irrigation or productive irrigation (blue)

Logical design combinations of parameters of operational objectives are grouped in stacks.

Choices: original design (bold font); new design (bold, italicized font); practiced (thick box border)

Figure 7.10. System and operational objectives, and flow control methods of Balanac RIS[1]

Parameters	Choices for design philosophy/system objectives			
Irrigation target	Protective		**Productive**	
Cropping system	**Rice mono cropping**		Diversified cropping	
Design irrigation season	**Wet season**		*Dry season*	
Irrigation during other season	Dry season: all area – fallow or small area – protective	**Dry season: all area – protective or small area – productive**	*Wet season: Supplementary*	
Irrigation supply in the system	Equitable supply to farmers	Equitable supply to hectares	**Flexible supply based on water need and availability**	Flexible supply based on water demand
	Choices for operational objectives of main system			
Decision-making on water allocation to tertiary unit (TA)	**Imposed allocation**		Semi-demand allocation	On-demand allocation
Method of water allocation to TA	Splitted flow	*Intermittent flow* / **Adjustable flow**	Adjustable or intermittent flow	Adjustable or intermittent flow
Method of water distribution through the main system	Splitted flow	*Rotational flow* / **Adjustable flow**	Adjustable flow	Adjustable flow
Flow control method	Proportional control	**Upstream control** or simultaneous control	Upstream control or predictive control	Downstream or volume control

Logical design choices for: protective irrigation (gray shade); productive irrigation (green shade); possible for either protective irrigation or productive irrigation (blue)

Logical design combinations of parameters of operational objectives are grouped in stacks.

Choices: original design (bold font); new design (bold, italicized font); practiced (thick box border)

Figure 7.11. System and operational objectives, and flow control methods of Sta. Maria RIS

A plausible logic for such low design water duty for productive irrigation was that the system was designed with the wet season as the main irrigation season. In other words, the irrigation was designed to supplement rainfall. This further implied that during the dry season, the system objective of productive irrigation was possible only for a portion of the total service area. Covering the whole service area would logically mean adopting protective irrigation. Otherwise, it was a case of over optimistic assumption on water requirements.

Meanwhile, the design operational objective of Sta. Maria RIS was the same as that of Balanac RIS as suggested by the same kind of adjustable gates of offtakes and cross regulators installed in the former. The original system design of Sta. Maria was logically coherent from the standpoint of design logic framework.

The information on recent developments and current practices on system operations, rehabilitation and improvements indicated some modifications in either the system objectives, operational objectives or both in the two NIS. In Balanac RIS, productive irrigation in the dry season remained one of the system objectives. This unchanged objective was shown in the adopted approach of formulating systems operation plans based on a reckoned optimum cropping calendar for rice production for the incoming crop year and on stipulated goals of system rehabilitation and improvement works of 100% service area coverage by irrigation. The fixed, long-crested proportional weirs constructed to replace the adjustable gates at the major flow distribution points along the main canal were an explicit sign of the change in the system objective of equitable supply to a unit area (ha) and amenability for flexible irrigation supply to limiting it to equitable supply. The Balanac RIS adopted the following changes in the parameters of operational objectives and flow control method: a shift in the method of water allocation to tertiary units and method of water distribution through the main system from flexible flow to splitted flow; reinforcement of imposed allocation; and shift from upstream control to proportional control. The reduction in opening of ungated offtakes based on their respective service areas was in line with the shift to proportional distribution of water.

Equitable flow per ha remained consistent with productive irrigation when applied in rice mono-cropping. While there was logical coherence among the parameters of the new operational objectives, they did not concur well with the system objective of productive irrigation. The fixed, proportional weirs that replaced the original gated offtakes, turnouts and cross regulators could simply divide any surplus or deficit in incoming canal flows in fixed ratio. They were not amenable for necessary adjustments in flow rate and volume that would be required to optimize crop growth and yield. The adoption of long-crested, proportional weirs was not well received by most farmers as evidenced by vandalized and chipped off crest length of all weirs constructed.

The present water distribution practice at major bifurcation points has a semblance with the adjustable flow method of water distribution that was intended in the original design, albeit without functional flow control structures. The present water allocation to tertiary units is through ungated culverts, open offtakes and rectangular weirs (Figure 7.12). The control of offtakes discharges and water level upstream is primarily accomplished by placing makeshift stoplogs such as bamboo and banana stems and wooden planks at the old cross regulators concrete structures (Figure 7.13).

Figure 7.12. Ungated circular turnouts (reduced culvert diameter at left photo) along the main canal of Balanac RIS

The current state of flow control structures along the main system of Balanac RIS will support neither splitted flow, rotational flow nor adjustable flow through the main system. The proportional weirs that would make a logical match for splitted flow were damaged while gates needed to achieve adjustable flow through the main system were non-existent. The ungated turnouts immediately downstream of proportional weirs but at lower elevation than the latter's crest level undermined fair and orderly water distribution. Similary, concrete structures of the original gated cross regulators enabled the use of improvised stoplogs immediately downstream of the duckbill or proportional weirs, effectively counteracting the latter. The incoherence among system and operational objectives, flow control structures and method and farmers preferences were highlighted by lack of functioning flow control structures and lower actual area irrigated, despite of higher water supply compared to Sta. Maria RIS.

Figure 7.13. Improvised cross regulators at distribution points officially redesigned for proportional flow along the main canal of Balanac RIS

While productive irrigation remained one of the system objectives in Sta. Maria RIS, its main irrigation season and irrigation supply were changed to the dry season and flexible supply, respectively. This view was deduced from the following management approaches: strategic seasonal adjustments in cropping calendar for different irrigation zones to meet the water requirements for each zone; a goal of 200% irrigation intensity in previous rehabilitation or system improvement projects; use of adjustable, gated flow control structures for distribution; and a 60% reduction in design service area. For the stated design diversion capacity of the system, this firmed-up service area of 974 ha translated to a water duty of 2.73 $l s^{-1} ha^{-1}$, which would be sufficient for productive irrigation during the dry season.

The water allocation to tertiary units or turnout service areas (TSA) remained an imposed allocation. The Sta. Maria RIS Irrigators Association (SANTAMASI) set the irrigation schedules in consultation with the TSA leaders or farmers' representatives prior to each cropping season. The present methods of water allocation to tertiary units of Sta. Maria RIS include adjustable flow, intermittent flow and splitted flow. Intermittent flow or on/off flow to minor tertiary offtakes has become the norm due to insufficient supply of water for simultaneous irrigation. It was achieved by using either adjustable gates or non-modular stopboards that were either fully open, passing the maximum flow or fully closed. Adjustable flow or varied flow was accomplished by adjusting the setting of the spindle-type gates and removing or adding pieces of modular stoplogs of tertiary offtakes. The splitted flow method of water allocation occurred in the case of ungated culverts and open offtakes.

The present water distribution through the main system of Sta. Maria RIS is either adjustable flow or rotational irrigation by secondary canals. The latter was more frequently implemented owing to the limited water supply for continuous irrigation. Water is diverted into the offtaking canals by maintaining the upstream water level at target levels by using adjustable or stepwise cross regulators.

The changes in the system objectives of Sta. Maria RIS logically conformed with the flow control method and in the changes in the methods of distribution through the main system and water allocation to tertiary units, except in the case of splitted flow that is achieved through ungated and open offtakes. Imposed allocation to tertiary units by adjustable flow and intermittent flow matches adjustable flow and either adjustable flow or rotational flow through the main system, respectively. These combinations are known to fit well with upstream control. Imposed allocation to tertiary units by splitted flow logically matches a splitted flow through the main system under protective irrigation.

The physical state of many flow control structures presented challenges in delivering the water as designed as many of the offtakes with adjustable gates are barely functioning; most needed replacement. Discharge and water level control structures were important for Sta. Maria RIS to enable to implement rotational irrigation, which is practiced during both wet and dry cropping seasons to cope with its limited water supply at the source. The tree trunks and leaves used as improvised gates for broken or missing offtakes and cross regulators (Figure 7.14) are vulnerable to unauthorized modifications. The many ungated, or open farm turnouts diverting water directly from the main canal compromised the integrity of the rotational irrigation as they can readily divert water meant for downstream farms. They disrupted orderly distribution of water and made management of water distribution more cumbersome.

The run-of-the-river dams of Balanac Dam and Sta. Maria Dam would not support well the systems objective of productive irrigation during the dry season. The pre-condition to designing an irrigation system with such objectives is that either water supply available at the source is adequate at any time, or sufficient water can be stored and made available to meet higher requirements associated with the dry season. The main purpose of run-of-the-river dams is to raise the water level at the source so that diversion by gravity is possible. They are not specifically designed to store irrigation water. Irrigation water supply through such dams is less reliable, making irrigated crop production vulnerable to fluctuations in river discharge. Reservoir-type dams would be a more logical match to irrigation systems designed for productive irrigation during the dry season, especially in the context of lower low river flow regimes attributed to climate change.

Figure 7.14. Improvised stoplogs as cross regulators along the main canal of Sta. Maria RIS

7.4 Initial rapid system diagnosis and performance assessment through RAP

The results of RAP computation of internal indicators of system performance based on the gathered information from system documents, interviews and walkthroughs for Balanac RIS and Sta. Maria RIS are summarized in Tables 7.5 to 7.10 and more detailed in Annex F.

In general, the values of the primary internal indicators of RAP were low, ranging from 0-2 (0 and 4 indicating least and most desirable, respectively). The only exceptions were in the case of travel time of a flow rate change through the main canals in both systems and in the case of control of cross-regulator along main canals of Sta. Maria RIS. The favourable ratings in travel time were attributed to the relatively shorter canal length of these two irrigation systems. The low values of indicators for water delivery, control and operation of flow control structures, and management and institutional setups of Balanac RIS and Sta. Maria RIS implied that the lackluster performances of these systems can be attributed to insufficiency or shortcomings in these aspects of water delivery structures and processes.

Table 7.5. Indicators of the water delivery services to farmers and by the main canal of Balanac RIS

	Actual	Stated
Service to individual ownership (i.e., field, farm)	1.1	2.0
• measurement of volumes	0.0	2.0
• flexibility	0.0	2.0
• reliability	1.0	2.0
• apparent equity	2.0	2.0
Service to field channels (most downstream point operated by paid employee)	0.9	1.9
• number of fields downstream of this point	0.0	1.0
• measurement of volumes	0.0	1.0
• flexibility	1.0	3.0
• reliability	1.0	2.0
• apparent equity	2.0	2.0
Service by the main canal to the second level canals	0.4	1.8
• flexibility	1.0	1.0
• reliability	0.0	2.0
• equity	1.0	2.0
• control of flow rates to the submain	0.0	2.0

The results of RAP indicate that Sta. Maria RIS was slightly better than Balanac RIS in terms of water delivery service by the main canal to its secondary canals, control and operation of offtakes and water level structures, canal condition, social order, WUA capability and budget. A more favourable evaluation results on water flow distribution along the main canals of Sta. Maria were attributed to the more functional cross regulators, offtakes and turnouts at its major distribution points. The stronger Sta. Maria WUA that was able to enforce the more defined rules on water distribution could be one of the reasons for lesser vandalisms and unauthorized water diversions and higher irrigation fees collection.

Table 7.6. Indicators of the control and operation of canals structures of Balanac RIS

	Main canal	Second-level canal	Third-level canal
Cross-regulator hardware in the canal	1.6	1.1	1.1
Travel time of flow rate change throughout the canal	4.0	--	--
Turnouts from the canal	0.7	1.3	1.3
Regulating reservoirs in the canal	0.0	0.0	0.0
Communications for the canal	1.0	1.3	1.3
Existence and frequency of remote monitoring at key spill points, including the end of canal	0.0	0.0	0.0
General conditions of the canal	1.4	1.2	1.2
Operation of the canal	1.3	1.3	0.8
Clarity and correctness of instructions to operators	1.3	1.3	1.3

Table 7.7. Indicators of the social order, water users, budget and employees of Balanac RIS

Social order in the canal system operated by paid employees	0.5
• deliveries are not taken when not allowed, or flow rates greater than allowed	1.0
• noticeable non-existence of unauthorized turnouts from canals	0.0
• lack of vandalism of structures	0.0
WUA	0.3
• percentage of WUAs who have functional/formal unit participating in water distribution	0.0
• actual ability of strong WUA to influence real-time water deliveries to the WUA	0.0
• ability of IA to rely on effective outside help for enforcement of its rules	0.0
• legal basis for the WUA	1.0
• financial strength of the WUA	1.0
Budgets	1.2
Employees	1.9

Table 7.8. Indicators of water delivery to farmers and by the main canals of Sta. Maria RIS

	Actual	Stated
Service to individual ownership (i.e., field, farm)	1.1	2.6
• measurement of volumes	0.0	3.0
• flexibility	0.0	3.0
• reliability	1.0	2.0
• apparent equity	2.0	3.0
Service to field channels (most downstream point operated by paid employee)	0.9	2.6
• number of fields downstream of this point	0.0	0.0
• measurement of volumes	0.0	2.0
• flexibility	1.0	3.0
• reliability	1.1	4.0
• apparent equity	2.0	2.0
Service by the main canal to the second level canals	1.2	3.0
• flexibility	1.0	3.0
• reliability	1.0	3.0
• equity	2.0	3.0
• control of flow rates to the submain	1.0	3.0

Table 7.9. Indicators of the control and operation of canals structures of Sta. Maria RIS

	Main canal	Second-level canal	Third-level canal
Cross-regulator hardware in the canal	3.1	1.9	--
Travel time of a flow rate change throughout the canal	4.0	--	--
Turnouts from the canal	2.4	1.7	--
Regulating reservoirs in the canal	0.0	0.0	--
Communications for the canal	1.5	1.3	--
Existence and frequency of remote monitoring at key spill points, including the end of canal	0.0	0.0	--
General conditions of the canal	1.4	1.8	--
Operation of the canal	1.9	1.9	--
Clarity and correctness of instructions to operators	1.3	1.3	--

Table 7.10. Indicators of the social order and water users, budget and employees of
Sta. Maria RIS

Social order in the canal system operated by paid employees	1.5
• deliveries are not taken when not allowed, or flow rates greater than allowed	2.0
• noticeable non-existence of unauthorized turnouts from canals	0.0
• lack of vandalism of structures	2.0
WUA	2.2
• percentage of WUAs who have functional/formal unit participating in water distribution	4.0
• actual ability of strong WUAs to influence real-time water deliveries	2.0
• ability of IA to rely on effective outside help for enforcement of its rules	0.0
• legal basis for the WUA	1.0
• financial strength of the WUA	1.0
Budgets	2.4
Employees	1.9

There were significant gaps between what the NIA system officials think they are doing in terms of water delivery services and the on-site assessments based on walkthroughs. The results of the system officials' and field assessments for Balanac RIS coincided only on the equity on water delivery service to individual fields and to the most downstream points operated by paid employees. In the case of Sta. Maria RIS, they only coincided on equity of delivery service to the most downstream points operated by paid employees.

7.5 Assessing system capacity and functionality

This section presents a comparison of the design, actual and required capacities of irrigation structures to perform their functions that are relevant to system operations (Tables 7.11 and 7.12). The design capacity was based on the information on the existing system configuration and design specifications of canals and flow control structures as stated in project documents. The actual capacity was estimated based on visual assessment of the on-site physical conditions and dimensions of canals including type, combinations and locations of flow control structures. The present capacity requirements were estimated based on the present service area and expressed need of system managers, operation staff and farmers. The system capacities that relate to the functions of irrigation structures relevant to canal operation are defined below:

- *diversion*. The capacity to divert the water from the main canal (parent canal) to the next lower-level canal (dependent canal) or to a delivery point the required rates, which can range from zero to maximum discharge capacity at this point;

- *division*. Refers to division of the total flow proportionally at key points, over a number of downstream outlets;

- *storage*. Relates to amenability of the irrigation system (canals or storage reservoir behind the dam or intermediate, regulating micro-reservoir) to impound water to be delivered at more convenient time and place or according to user's requirement;

- *conveyance*. Relates to adequacy of canal dimensions and linings to, as the term implies, convey the required irrigation water along the canal reach;

- *sediment control*. Pertains to mechanisms of removing excessive sedimentation from the canals and flow control structures;

- *discharge transfer*. Refers to the time lag between a change of discharge upstream and its conversion at a downstream point of the canal;

- *water level control*. Is about stabilizing the required water level (head) in the parent canal;

- *flow measurement*. Relates to the accuracy and reliability of its flow measurement structures at strategic locations (e.g.: critical bifurcation point, offtake);

- *safety*. Pertains to magnitude of freeboard and presence of escapes for disposal of spill water at critical points;

- *communication*. Relates to availability of real-time information important to system operations;

- *water reuse*. Relates to use of structures to recapture irrigation water along drainage and natural water ways;

- *transport and roads*. Pertains to the accessibility of and travel time to the whole physical structure for purposes of supervision of system operations and maintenance or urgent repairs.

Table 7.11. Physical capacities of Balanac RIS vis-a-vis the design capacity and capacity requirement

Capacity	Capacity aspect	Actual versus design	Design versus requirement
1. Division	Dysfunctional duckbill weirs, ungated direct offtakes	<	<
2. Storage (canal)	No online mini reservoirs; canal wall seepage; big canals	≃	<
3. Conveyance	Cracks on canal linings; siltation; direct, ungated offtakes and drainage inlets; rubbish dumps	<	<
4. Sediment control	Dysfunctional dam sluice gates; silt entry over canal berms	<	<
5. Discharge transfer	Within the day	≃	≃
6. Water level control	Dysfunctional long-crested weirs	<	<
7. Flow measurement	Not done; missing staff gauge	<	<
8. Safety	With freeboard; no spill points except end checks	≃	<
9. Communication	Mobile phones	>	<
10. Water reuse	7 check gates	≃	<
11. Transport/access	20%	<	<
12. Diversion (dam)	Minimal head	<	<
13. Storage (dam)	No storage; heavily silted	<	<

> increased   decreased   ≃ same

The results of the comparison between the actual and design and between the design and actual requirement suggest that Balanac RIS and Sta. Maria RIS had either decreased capacity or inherently limited capacity to perform the system various functions and to satisfy present demands.

The flow cross-sectional areas of the existing canals of Balanac RIS were adequate to convey their design discharges, except for its fourth-level canals. Lateral A1A, which branches out from Lateral A1, has a service area of 172 ha, or about two-thirds of the service area of the latter and design conveyance capacity of 0.15 m^3s^{-1}. Based on its service area and the NIA standard design values, its conveyance capacity should have been at least 0.26 m^3s^{-1}. In Sta. Maria RIS, there was no problem conveying the design flow through its canals.

Table 7.12. Physical capacities of Sta. Maria RIS vis-a-vis the design capacity and capacity requirement

Capacity	Capacity aspect	Actual versus design	Design versus requirement
1. Diversion (canal structures)	Some dysfunctional offtakes and cross regulators	<	~
2. Storage (canal structures)	No online mini reservoirs; big canals	~	<
3. Conveyance	Sediment deposits; direct ungated offtakes and drainage inlets	<	~
4. Sediment control	Dysfunctional sluice gates; silt entry over canal berms	<	<
5. Discharge transfer	Within the day	~	~
6. Water level control	Some modular stoplogs and damaged cross regulators	<	~
7. Flow measurement	Silted/influenced flume approach; missing staff gauge	<	<
8. Safety	With freeboard; no spill points	~	<
9. Communication	Mobile phones	>	<
10. Water reuse	9 check gates	~	<
11. Transport/access	92%	~	~
12. Diversion (headwork)	Minimal head; intake gate spindles askew	<	<
13. Storage (headwork)	Silted dam	<	<

> increased   decreased   ~ practically the same

The concrete or grouted riprap linings improved the conveyance and transfer capacities of the main and second-level canals of the two systems. All canals were originally all earthen or unlined. At present about 50% of the main canals and second-level canals of Balanac RIS and 71% of that in Sta. Maria RIS are lined with either concrete, grouted riprap or concrete hollow blocks. These linings resulted in higher flow velocities, hence reduced the time lag between a change of discharge upstream and its conversion at downstream points of the canal.

However, the overall improvements of conveyance capacities and transfer capacities, which were also attested by canal operators and farmers, were constrained by noticeable siltation and weed growth in a number of stretches of the unlined canals of the two systems. There were also seepage losses through lined canal embankments along a number of main canal reaches of Balanac RIS (Figure 7.15).

Figure 7.15. Balanac RIS main canal reach with considerable grass growth (upper bank) and seepage loss (lower bank)

Balanac RIS shifted from upstream control to the proportional control system. With significant segments of all proportional weirs destroyed, except for Lateral A bifurcation, both the intended division and water level control capacities at major bifurcation points of the Balanac RIS were practically obliterated (Figure 7.16). The discharges across the open, or ungated offtakes at these junctions were relatively more sensitive to water level fluctuations in the main canal, or parent canal due to the weir flow characteristics. Diversions of water through these offtakes, including Lateral A offtake that has the long crested weir still intact, were effected by using tree trunks and branches as cross regulators. This manner of water diversions resulted in unruly water distribution.

In Sta. Maria RIS, the main canal orifice-type offtakes and major turnouts have lower canal beds, or floors at their downstream sides enabling free-flow conditions, hence, good and controlled diversion capacity for the dependent canals. A super-critical flow through the offtakes and turnouts ensured the independence of discharge into the second level canal from its tailwater conditions. It ascertained no higher discharge than what is intended to be diverted. Further, the orifice-flow offtakes were less sensitive to water level fluctuations of the main canal and, thus, were capable of maintaining relatively constant discharge into their dependent canals. However, a number of missing, or non-operational vertical gates of these structures and the ungated culverts compromised the integrity of the distribution capacity of the irrigation system (Figure 7.17).

(a)   (b)

(c)   (d)

Figure 7.16. Water division structures at major bifurcation points in Balanac RIS: Buboy-main canal (a); Lateral A-main canal (b); Laterals A-A1 (c); Laterals A1-A1A (d)

Spindle-type vertical gates were used as cross regulators along Sta. Maria RIS main canals and Lateral A, a major second-level canal. Twenty one of these cross regulators were procured and installed under a system improvement component of a foreign-assisted project in 2011. Being an orifice-type structure or having an undershot-flow, these vertical gates were relatively more sensitive to variations in the water level than overshot structures. Thus, they were less suitable for water level regulation. An apparent social acceptance of these vertical gates in Sta. Maria RIS may be due to their amenability for rotational irrigation being practiced to cope with the relatively low water supply.

Figure 7.17. Inoperable (spindle and frame hit and bended by a wayward truck) Lateral A offtake of Sta. Maria RIS

There were neither river gauging activities nor measurements of diverted water by the main intake structure and by offtakes at major bifurcation points being carried out for Balanac RIS. There were also no river gauging activities for Sta. Maria RIS. Though the water diverted for the irrigation is measured through the Parshall flumes by its irrigators association, discharge estimates based on these water level readings would be unreliable because the flumes were either with considerable siltation at the floor or at the canal approach side. There were flumes without the staff gauges at the upstream end of each second-level canal of Sta. Maria RIS, but they were not used.

The major distributary or main system canals of Balanac RIS and Sta. Maria RIS have enough freeboard. However, there were no escapes for disposal of excess water along the main canals of each system, except at the end of these canals. Their tailwaters drain ultimately into the Laguna Lake. They both suffered seasonal submergence by the Lake during the monsoon months.

The main canal that is fed by Bagumbayan Dam of Sta. Maria RIS can divert water to the system's second main canal (Mata Main Canal) through a connecting canal. Such diversions were only done to supplement water supply for the service area of the latter in rare occasions that the former can share some of its water.

Check gates were constructed in Balanac and Sta. Maria RIS in recent years for the purpose of recapturing drainage water of upstream farms and diverting such water into downstream farms (Figures 7.18 and 7.19). These check gates were irrigating about 21% and 23% of the firmed-up service areas of Balanac RIS and Sta. Maria RIS, respectively (Table 7.13).

Figure 7.18. Check structures used to recapture drainage and tap additional water in
Balanac RIS

Communication of water delivery concerns and operation instructions among system officials and field personnel have dramatically improved with the full coverage of wireless telecommunication networks in the service areas and use of mobile phones by managers, canal and gate operators of Balanac and Sta. Maria RIS. Communication allowance for official calls and text messaging were given to operations staff out of the O&M budget.

Cabooan

Pague

Indocina

Samala

Tagalag

Cambuja

Figure 7.19. Check structures used to recapture drainage and tap additional water in Sta.
Maria RIS

There is an access road passable by a 4-wheel vehicle to about 60% of the Balanac RIS Main Canal while the remaining length is accessible only by foot. The access road is located only on one side of the canal. Considerable stretches of the access road to the main canal in the downstream area were occupied by informal settlers. While Buboy Extension has an access road on one of its sides, Lateral A and its two sub-laterals either did not have access roads or informal settlements had encroached on them. The operator of the main intake structure and the canal operators were using a motorcyle and bicycles as service vehicles.

Table 7.13. Check structure to capture and divert drainage and additional water

Check gates	Reused water	Imported water	Total area
Balanac RIS			
• Biñan	93.7	0.0	97.3
• Salasad	108.6	0.0	108.6
• Total area	212.3	0.0	212.3
Sta. Maria RIS			
• Cabooan	38.4	0.0	38.4
• Pague	28.5	0.0	28.5
• Masinao and Samala	29.2	0.0	29.2
• Indocina	11.6	29.9	41.5
• Tagalag	0.0	45.2	45.2
• Cambuja	48.3	0.0	48.3
• Total area	156.0	75.0	231.0

Almost the whole length of Sta. Maria Main Canal and its second-level canals, except Lateral E, have access roads on one side of the respective canals. While these access roads are passable by 4-wheel vehicle, access to the canal sections were constrained by small trees (mostly citrus specie) and vegetable shrubs planted in the narrow strip between the roads and the canals. Motorcycles were used as service vehicle of system operators of Sta. Maria RIS.

Balanac RIS and Sta. Maria RIS have no significant storage capacity owing to the nature of their run-of-the-river headworks. The heavy silt accumulations at the upstream side of Balanac Dam and Sta. Maria dams reduced the capacity of the headworks to divert water (Figure 7.20). It also eliminated the inherently little capacity of the system to regulate the fluctuations in the diverted water. There is no intermediate storage within their service areas. The dysfunctional sluice gates at the headworks removed the capacity to minimize or prevent sediments entering the main canals.

(a)

(b)

Figure 7.20. Sediment deposits upstream of Balanac Dam (a) and Bagumbayan dams
with permanently closed sluice gate (b)

7.6　Sensitivity analysis

The study intended to analyze the behaviour of Balanac RIS and Sta. Maria RIS under their respective typical water distribution practices through the assessment of the sensitivity of their irrigation structures at three levels: on each main type of flow control structure (e.g.: offtake, water level regulator) taken in isolation; at key diversion and division points; and along canal reaches. The proposed research method was to evaluate the sensitivity from water level and discharge records and through direct field measurements.

Sensitivity assessment by using records was not possible in the case of Balanac and Sta. Maria RIs since flow measurements are not carried out at key distribution points along the canal networks of the two systems. Meanwhile, direct field measurements proved to be unwieldy, or intrusive. The current practice of open ended cropping seasons and unrestrained offtaking along the bifurcating and ungated canal network of Balanac RIS made it practically impossible to maintain flow conditions required for the sensitivity experiments. While there were water delivery schedules for the second-level canals and key distribution points through mostly gated offtakes of Sta. Maria RIS, its water distribution practice did not allow manipulation of flow for the purposes of sensitivity measurements. It follows tight schedules of rotational irrigation especially during the dry cropping seasons. During the rainy months, the main intake structure is closed to prevent entry of flood water through the canal network and some ungated offtakes, hence, submergence of the cropped areas.

The assessment of sensitivity through field observations on the types and conditions of flow control structures and visualisation of relative flow changes based on their respective discharge-head relations was resorted to as the pragmatic approach in the case of Balanac RIS and Sta. Maria RIS. From the generic flow equation (equation 5.5) and its logarithmic derivative (equation 5.6), an overshot offtake ($\alpha = 1.5$) is more sensitive than an undershoot offtake ($\alpha = 0.5$) for a given water level variation (Table 7.14). Based on the same equations, an overshot cross regulator is less sensitive to head variation than an undershot cross regulator. Inherent to their hydraulic characteristics, undershot offtake and overshot cross regulator are better suited for discharge and water level regulators, respectively.

Also the diversion structures at major distribution points were designated a diversion sensitivity, or hydraulic flexibility classification according to the resulting variations in discharges of the structures in response to variations in incoming flow as conceptualized by Horst (1998) and Ankum (2001). In this conceptualization of hydraulic behaviour of diversion structures, a hydraulic flexibility, or a ratio of discharge variation in offtaking canal to discharge variation in the parent canal equal to unity ($F = 1$) is conceived to have proportional distribution of flow fluctuations through the canal network. Low sensitivity diversion structures ($F < 1$) are considered to pass incoming flow fluctuations to the tail-end of the canal. In case of high sensitivity diversion structures ($F > 1$) the effect of incoming fluctuations is manifested the most in the upstream canal reach. The combinations of diversion structures that yield an F = 1 is recommended by both authors for adoption in irrigation systems under upstream control.

Table 7.14. Indicative sensitivity at branching points of Balanac and Sta. Maria RIS

Main branching points	Offtake/turnout	Cross regulator	Hydraulic flexibility F
Balanac RIS			
MC - Lat. Buboy	Overshot	*Overshot[1]*	F > 1
	Undershot		F < 1
MC - Lat. A	Overshot	Overshot	F ≅ 1
	Undershot		F < 1
Lat. A - Lat. A1	Undershot	None	F < 1
Lat. A1 - Lat. A1A	Open	None	F > 1
Sta. Maria RIS			
SMMC - Lat. A	*Undershot*	Undershot	F ≥ 1
SMMC - Lat. B	Undershot	Undershot	F ≥ 1
SMMC - Lat. C	Undershot	Undershot	F ≥ 1
SMMC - Lat. D	Undershot	Undershot	F ≥ 1
MMC - Lat. E	Undershot	Undershot	F ≥ 1
MMC - Lat. F	Undershot	Undershot	F ≥ 1

[1]In italics means non-functional structure

The diversion structures in Balanac RIS and Sta. Maria RIS were designed to proportionally distribute any changes in canal discharge among the offtakes (F = 1). This gate proportional diversion objective was surmised based on the combination of the type of installed discharge and water level regulators and the generally accepted principle of equitable flow distribution to ha. In general, the gate-proportional diversion in upstream control is approximated by a combination of either both overflow or both underflow discharge and water level control structures.

The lack of functional flow control structures at the bifurcation points of Balanac RIS, except at the MC-Lat. A, suggested unproportional diversions at these points due to either higher or lower sensitivity at the offtaking canal than that of the parent canal. For a change in incoming flow, the lack of regulation structures would result in a larger discharge variation at the offtaking canal than in the continuing canal and other downstream offtakes without flow control structures. In the case of Lateral A - Lateral A1 junction, the offtake (Lateral A1) has lower sensitivity owing to its hydraulic characteristics (discharge-head relation). It takes unproportional lower discharge variation than the ongoing canal (Lat A). Proportional weirs for the continuing Balanac Main Canal and offtaking canal Lat. A was aimed for proportional flow distribution at this junction (Figure 7.21). However, this objective was altered by placing stoplogs at the concrete structure of the original adjustable water level regulator. Such alteration was easily done at Buboy and other bifurcation of Balanac RIS due to the groove on the upstream side of each concrete structure.

Figure 7.21. Proportional weirs at MC-Lat. A bifurcation in Balanac RIS

A combination of open offtake and undershot cross regulator (vertical gate) at the junction Sta. Maria Main Canal (SMMC) and Lat A would result in higher discharge variations in the offtaking canal for small fluctuations in the incoming flow (F > 1). In theory, gate proportional diversion (F = 1) can be achieved at branching points with both undershot flow control structures at offtaking and main canals by adjusting the gates according to the changes in the incoming water supply. However, in the case of Sta. Maria RIS, where the water supply can be highly variable, operating the system to achieve gate proportional diversion (F = 1) is too cumbersome and impractical. The offtaking canal would have higher discharge variation than the continuing canal (F > 1).

7.7 Origins and causes of perturbations

The research aimed to identify the perturbations of discharge and water levels along main canals in terms of origin, magnitude, timing and frequency in order to know whether the main canals are self-reacting, or if specific interventions (e.g.: adjustments in system operations and management or changes in flow control structures) must be carried out to maintain the water level and discharge at the farm offtake/turn out within a practically reasonable range of the target levels. The results of field investigations suggested that the causes of perturbation included unauthorized direct offtaking from main canals, irrigation return flows and overland flows entering the main canals, fluctuating water supply from the river and unregulated diversion, lack of flow control structures and checking, or flow obstruction by the locals for domestic purposes (livestock tending, laundry and swimming).

By design, direct offtakings from main canals through official, ungated, or open turnouts have been practiced in both Balanac and Sta. Maria RIS. Results of interviews showed that direct offtaking through unofficial turnouts have been tolerated to either facilitate, or effect delivery of water for groups of farms. Laissez faire cropping schedules, non-functional flow control structures and uncoordinated offtakings contributed in lower predictability of water level and discharge fluctuations in Balanac RIS.

There was no lull in irrigation activities along the main canals in both systems as upstream farmers in Balanac try to have a third cropping and tight rotational irrigation is followed in Sta. Maria RIS, except during irrigation cutoff to give way for construction works along the canal, or very low flows from the water source. The study on perturbations was limited to field identification of the origins and causes of perturbations owing to these situations. Locations of direct offtakes and drainage inlets along the main canals are summarized in Table 7.15 (Annex E).

7.8 Conclusions

There was logical coherence among the design philosophy, or overall system objectives, operational objectives and on-canal flow control structures of the original system design of

Balanac RIS and Sta. Maria RIS from the view point of the design logic framework. The design philosophy and system objective of productive irrigation for rice mono-cropping was in support to the rice self-sufficiency goal of the country. The use of gated flow control structures is consistent with the need for an adjustable flow method of water allocation and distribution along major conveyance canals. Adjustable flow was in turn a logical design choice to achieve flexible supply that fits well with the productive irrigation objective.

Table 7.15. Count of direct offtakes and ungated turnouts

	Open/stoplog	Unauthorized
Balanac		
Main canal	118	73
Lat. Buboy	22	2
Lat. A	39	19
Lat. A1	19	10
Lat. A1A	23	7
Sta. Maria		
SMMC	43	20
MMC	68	30
Lat. A	69	24
Lat. B	9	2
Lat. C	15	1
Lat. D	4	2
Lat. E	15	3
Lat. F	29	0

However, the shift to splitted flow and proportional water distribution and flow control methods, respectively, in Balanac RIS did not logically agree with its unchanged system objective when subjected to the same design logic analysis. The results of RAP carried out in Balanac RIS supported the hypothesis of unwieldy water distribution deduced from the design logic framework analysis. The splitted flows achieved by using the fixed proportional structures did not sit well with farmers as manifested by vandalisms that rendered these structures non-functional. The values of RAP internal indicators for water delivery, control and operation of flow structures and social order were low, mainly due to the lack of functional flow control structures and presence of ungated direct offtakes and turnouts.

From the design logic perspective, a number of recent system improvements and present system operations in Sta. Maria were consistent with the shift in its system objectives of productive irrigation from the wet season as the main irrigation season to the dry season. Despite this coherence in the general system design and operations, the water delivery service was rated low in RAP context due to the poorly functioning gates and presence of open turnouts or offtakes.

The capacity of the structures of Balanac RIS and Sta. Maria RIS to perform their intended functions had decreased due to damage, defects, dysfunctions, missing parts and deviations from preconditions for proper functioning through time. Repairs, or replacements are required for sluice gates of the dam, main intake gates and discharge and water level control structures. Maintenance work to restore head differential (diversion capacity) at the dam and approach conditions (flow measurement capacity) for the flumes are necessary.

In Balanac RIS, the most telling capacity issue was the division capacity. The social acceptability of proportional, or splitted flow distribution and the technical soundness of the design configuration of division structures (proportional long crested weirs and associated offtakes) must be revalidated. The most striking issue in Sta. Maria RIS was the limited water supply from the rivers. Options to increase the storage capacity of Sta. Maria dams or augment the water supply from other sources must be investigated.

Quantifying sensitivity proved to be impractical in the case of Balanac RIS and Sta. Maria RIS where the water level is uncontrolled and ungated, direct offtakings were not managed. A combination of ocular inspection and general hydraulic characteristics (disharge-head relation) of flow control structures served as a pragmatic tool to gauge the sensitivity of the adopted design or present on-site configuration of the flow control regulators. The conceptualized diversion sensitivity or hydraulic flexibility of commonly used design configurations of diversion structures (combinations of offtake and water level controls) is a sufficient alternative for the purpose of diagnostic assessment.

The combination of the logic design framework and the adaptively modified RAP and capacity and sensitivity assessments of the physical structures provided a fundamental structure of a system diagnostic approach for a critical examination of the root causes of underperformance of Balanac RIS and Sta. Maria RIS. The design logic framework brought the assessment of system performance through RAP in perspective with the design philosophy, overall system objectives, operational objectives and physical features of flow control structures. It provided a clear framework on troubleshooting any inherent inconsistency in system design during the planning stage. It clarified the benchmark where the system water delivery performance can be fairly judged. Meanwhile, the RAP compliments the capacity and sensitivity assessments.

The present logic design framework and diagnostic tools of MASSCOTE focus on canal structures design and operation. Inclusion of the headwork component in both diagnostic approaches will put the water supply issues arising from climate change in clear perspective. With a prolonged, drier dry season, systems designed with a productive irrigation objective, but without a storage-type dam are facing chances of failure. Such combination of diagnostic tools offers a comprehensive system diagnosis approach with system modernization orientation, particularly suitable for mostly ungauged, run-of-the-river type irrigation systems in the country.

8. REVALIDATION OF DESIGN ASSUMPTIONS ON PERCOLATION AND WATER SUPPLY

8.1 Introduction

Overoptimistic design values and assumptions in water supply and irrigation efficiencies has been said to be a major reason for the significant gap between the design service area and actual area irrigated of most canal irrigation systems in the country. Overestimation of dependable supply and underestimation of seepage and percolation would result in larger design service area than what can be possibly irrigated with the given water supply. Miscalculations of design values could be due to broad generalization of data input in the face of limited information, expediency for project implementation and use of inadequate methods in estimating available water supplies. Further, changes in land use in the basin and in weather patterns, which are known to influence the river hydrographs, compounds the uncertainty in the validity of the values of design parameters.

In the above cases, revalidation of design parameters forms part of a prudent approach to the formulation of irrigation modernisation plans. The findings of the study of David *et al.* (2012b) highlighted the importance of using site-specific values on percolation. The results of the field measurements of percolation suggested that much lower percolation rates used during the design stage of nine national irrigation systems (NIS) resulted in gross underestimation of irrigation water requirement and, consequently, design discharge capacities. In the case of Balanac RIS and Sta. Maria RIS, operation engineers adopted the percolation values found in old project documents in preparing the system operation and management (O&M) plan. Design engineers seldom concern themselves with water balance parameters such as percolation since most of detailed engineering design works only involve restoration and improvement of existing physical structures.

Rainfall-runoff relations for most river basins feeding the canal irrigation systems in the country are not yet well studied. Many of the rivers tapped as water sources for canal irrigation systems remained ungauged. Flow gauging activities were even discontinued in some rivers feeding NIS. These stopped in 1979 for Balanac River, while there has been no gauging station in Sta. Maria River basin.

The dependable water supply for the irrigation systems can be expected to have changed with changes in land use of the river basins. The effects of changes in dependable flows are more readily felt in these run-off-the-river diversion irrigation systems than in storage, or reservoir type systems. How the recent changes affecting the water supply sources were accounted for in the estimation of dependable flows reflect the accuracy of the estimates, hence, their usefulness to system operation and modernisation project planning. While streamflow gauging has started in a number of irrigation systems and plans to collect such baseline information are gaining support in recent years, a pragmatic approach to sense the adequacy of river discharges for irrigation improvement projects in the immediate future would need to be formulated.

8.2   Research methodology

In this study, field measurements to estimate, percolation and irrigation conveyance efficiencies were carried out within the service areas of Balanac RIS and Sta. Maria RIS. Measurements of percolation and farm ditch conveyance losses were done using the 3-cylinder method and ponding test, respectively. The ponding test relates conveyance loss rate to the drop in water level in an initially water-filled, isolated canal section (no inflow and outflow occurring), taking into account the evaporation. Similarly, percolation is determined by relating the changes in water depths in a set of three cylinders, one with water-tight bottom and two without bottoms, to evaporation alone, evaporation and percolation or a combination of these three parameters. The changes in water level of each cylinder and rain gauge reading were algebraically added to estimate the water that percolated.

These direct physical measurement techniques have been the standard method for percolation and seepage measurements. Canal seepage has also been traditionally estimated by using other direct measurement techniques such as seepage meters, and inflow-outflow tests; indirect methods like monitoring the groundwater table adjacent to the canal; and by prediction using empirical or analytic formulae. The ponding test was selected to estimate the canal conveyance losses owing to the attributed better level of accuracy of this particular technique when applied to small canals.

A collaborative work to identify the texture of the soil of the experiment sites through the hydrometer method was carried out with students as part of the latter's thesis requirement for Bachelor degree. The corresponding percolation values for the identified soil texture of the sites where determined from the design textbook and manuals. They were compared to the measured percolation values.

To get a sense of the adequacy of the river flows available for irrigation systems, local knowledge on historical flows was gathered through informal interviews and field visits to the area during both cropping seasons throughout the course of the study. Old residents and long-time farmers, including members of irrigators associations were asked on the sufficiency of diverted water in relation to their irrigation need and observed river flows. Available water flow records were sought from water users associations. Also, sample canal discharge measurements during dry seasons were carried out.

The available flow data and basin characteristics of the rivers feeding the Balanac RIS and Sta. Maria RIS were collated and/or generated. Appropriate methods that can be used to estimate river flows at main diversion or dam sites were identified based on the availability of data input, applicability to the case study and practicality of computations.

8.3   Field measurements of percolation and farm ditch losses

The number of experiment sites for percolation and ponding tests were determined in such a way that there was at least one for each major soil type, relative to location (upstream, midstream and downstream) and every 100 ha of the service area. For each identified experiment area, a representative farm ditch and a paddy field served by the former were selected as experiment site. The considerations for the selection of the exact sites included proximity to a water source for a better control of the water level within the paddy field, least risk to vandalisms and willingness of farmers to accommodate the experiment. Nine and 10 experiment sites were selected for Balanac RIS and Sta. Maria RIS (Figures 8.1 and 8.2).

The field measurements of percolation and conveyance losses were carried out during May-June 2013 and 2014 in Balanac RIS and from September-December 2012 in Sta. Maria RIS. These experiments were done during lull in farming activities in between the cropping seasons, while the experiment sites and surrounding fields were saturated with water and isolation of sections of canal was allowed. As such the field experiments were coordinated with farmers, irrigators association and NIA to avoid conflicts with their normal irrigation and on-farm water management activities.

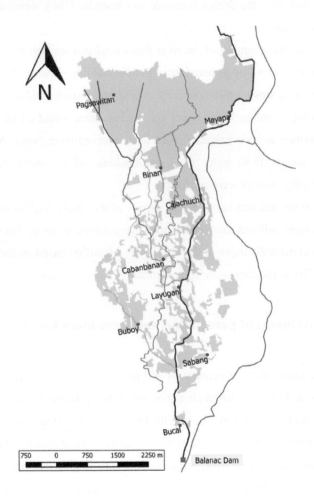

Figure 8.1. Sites of percolation and ponding experiments in Balanac RIS

A set of three 40-cm diameter galvanized iron (GI) cylinders and a GI storage-type, standard rain gauge were installed in the selected paddy fields (Figure 8.3). The set of cylinders consisted of a 50-cm high without bottom ("P" cylinder), a 40-cm high with a water-tight bottom ("E" cylinder), and a 30-cm high without bottom ("S" cylinder). Each cylinder had a 3-inch diameter stilling basin welded at its inside wall. The "P" cylinder was driven until its lower end touched the plough pan of the paddy field. The "E" cylinder was lowered into the hole dug into the paddy field up to a depth that allowed it to have the same length of exposed walls as of the "P" cylinder. The excavated soil was placed inside the "E"cylinder. The "S" cylinder was driven into the soil until the same length of the exposed wall was achieved and its lower end did not touch the hard pan. At least a metre distance from the levee and between cylinders was observed for ease of the measurement readings and to minimize errors due to possible shading from the sun's rays and shielding from the wind of the cylinder by the levee. The rims of the cylinders were levelled by using a carpenter's level bar. The same water level was maintained inside and outside the cylinders to minimize thermal mismatch errors.

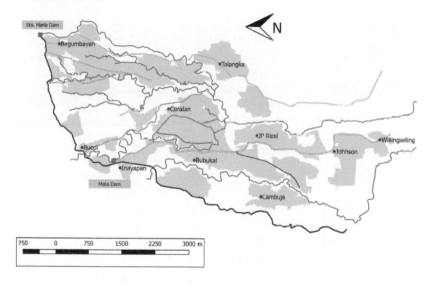

Figure 8.2. Sites of percolation and ponding experiments in Sta. Maria RIS

The water levels were measured inside the stilling basin by using a hook gauge. The water level readings for the "E" cylinder measured evaporation, while those of "P" cylinder measured the combined evaporation and percolation. The difference in their readings was taken as the percolation loss. When adjusted with the evaporation and percolation measurements obtained from "E" and "P" cylinders, the readings on the "S" cylinder would give the estimate of the seepage loss (Figure 8.4). The measured pan evaporations were adjusted by a factor of 0.7 to account for pan coefficients or evaporation associated with metal pans. The pan coefficient of 0.7 was selected based on moderate wind speed of about 2.5 m s^{-1} and a high relative humidity greater than 70% observed at the nearest PAGASA weather station in UP Los Baños. The rain gauge readings were used to correct the measured water levels in the cylinders for any rainfall that occurred.

Figure 8.3. Three-cylinders method for field measurement of percolation in Balanac RIS

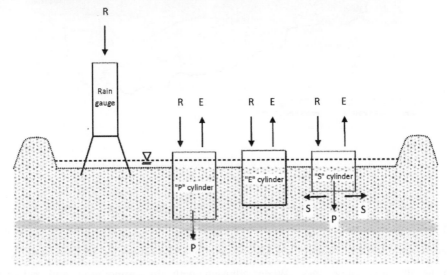

R - rainfall; E - evaporation; P - percolation; S - seepage

Figure 8.4. Scheme of the three-cylinders method for field measurement of percolation

At least two measurements of water levels in the cylinders each day, one in the morning and one in the afternoon, were made. Additional measurements of water levels in the cylinders were done in the event of high percolation and seepage rates to avoid water level dropping too low or, in the event of heavy rainfall, to avoid overflowing of water from the cylinders. Rainfall measurements were done either immediately right after the rain as the collected rainfall might evaporate quickly or just before an imminent rain occurrence. For each site, measurements were continued until three weeks or until the readings stabilized.

Ponding tests were carried out on a section of each farm ditch serving the paddy field where percolation measurements were conducted simultaneously (Figure 8.5). The selected canal sections were more or less straight, with uniform cross section, at least 20 metres in length with well-maintained banks. They were isolated by installing a rectangular GI sheet across its two ends, thus, creating a pond. A plastic-sealed earth dike or mud embankment at least a metre from each end of the isolated canal section was constructed to create a buffer zone. A perforated, 20-cm diameter GI sheet stilling basin was installed at the middle of the isolated or ponded canal section for measuring the water level in the canal. Its surface was levelled by using a carpenter's level bar.

Figure 8.5. Ponding test setup in Balanac RIS

At the start of each ponding test run, the isolated section and the buffer ponds were filled with water up to the designed full supply level of the canal at least 12 hours before measuring the water surface drop. The measurements of the water level and top width of the water surface at one-meter interval in the ponded section were taken at least twice a day. The water level was allowed to drop within the reading range capacity of the hook gauge. Just before the readable maximum drop is reached, the pond was refilled with water and another round of measurements started. The measurements were continued for about three weeks or until the readings stabilized, or when at least three consecutive readings were comparable. They were corrected for observed rainfall.

8.3.1   Results of the percolation tests

The adjusted evaporation values averaged from 2-4 mm day^{-1} in the two systems. These values were reasonably in the same order of magnitude of the values observed at the National Agrometeorology Station of PAGASA during the same period and the International Rice Research Institute (IRRI) in Los Baños, the two nearest weather stations.

Percolation rates in Balanac RIS averaged from 1-30 mm day^{-1} (Table 8.1 and Figure 8.6). Only four of the nine experiment sites (Sabang, Pagsawitan, Calachuchi and Mayapa) had average percolation rates within the assumed value for the system of 2 mm day^{-1}. Two of the nine sites (Layugan and Cabanbanan) had average values greater than 28 mm day^{-1}. Notably, the seepage rates observed in these sites were equally high as their respective percolation rates. Another two sites (Buboy and Bucal) had significantly different average percolation rates for two distinct soil conditions. From the relatively settled soil of the experiment site during the first half of the duration of the experiment to disturbed, surrounding soils as a result of second ploughing, the percolation values for Buboy increased from 5 mm day^{-1} to 22 mm day^{-1}. Meanwhile, the average percolation rates for Bucal experiment site decreased from about 12 mm day^{-1} to less than 1 mm day^{-1} when adjacent fields became saturated. These changes in percolation values following the respective changes in field conditions were generally expected. There were zero and small negative values of percolation measured in the Balanac experiment areas, except for Layugan, Cabanbanan and Buboy experiment sites. These values could be due to upswelling of seepage flow.

Table 8.1. Results of percolation tests (mm day^{-1}) in Balanac RIS

Site	Location	Soil condition		P Avg	P Min	P Max	S Avg	S Min	S Max	E Avg	E Min	E Max
Bucal	N 14° 12' 32.3"	Fallow; soaked for ploughing;	PMAM	12.7	6.0	19.7	40.2	5.9	78.1	2.2	1.6	2.7
	E 121° 26' 24.8"	unsaturated adjacent fields	AMPM	12.2	3.7	19.2	45.0	8.9	81.0	2.5	1.5	3.8
		Fallow; soaked for ploughing;	PMAM	1.3	0.0	3.7	50.9	21.9	92.8	2.1	1.2	3.5
		saturated adjacent fields	AMPM	1.0	-1.8	2.7	47.8	29.7	74.1	2.2	1.0	3.5
		Full duration	PMAM	5.1	0.0	19.7	46.7	5.9	92.8	2.1	1.2	3.5
			AMPM	4.7	-1.8	19.2	46.7	8.9	81.0	2.2	3.8	1.0
Sabang	N 14° 13' 28.1"	Settled; land soaking	PMAM	1.6	-0.5	7.9	3.1	-1.8	12.9	3.8	1.5	6.7
	E 121° 26' 47.5"		AMPM	1.5	-0.8	4.9	2.9	-1.1	6.6	3.7	2.0	6.4
Buboy	N 14° 13' 52.6"	Settled; land soaking	PMAM	5.5	4.5	8.8	17.4	3.7	29.0	2.1	1.0	2.8
	E 121° 25' 50.6"		AMPM	5.3	3.4	8.5	17.7	6.1	31.1	2.2	0.5	3.6
		Disturbed; after second	PMAM	22.3	11.1	39.4	6.0	-13.6	34.7	4.0	2.6	6.4
		ploughing	AMPM	22.1	10.5	36.9	2.8	-33.5	29.9	3.8	6.9	0.6
		Full duration	PMAM	15.6	4.5	39.4	10.4	-13.6	34.7	3.2	1.0	6.4
			AMPM	15.7	3.4	36.9	8.9	-33.5	31.1	3.1	0.5	6.9
Layugan	N 14° 14' 22.2"	Newly transplanted	PMAM	30.1	8.0	54.6	14.4	-12.3	59.7	3.5	1.7	5.7
	E 121° 26' 24.2"		AMPM	32.3	7.9	56.0	12.9	-15.6	54.0	3.4	1.6	6.0

Modernisation strategy for the national irrigation systems in the Philippines

Table 8.1. *Continued*

Site	Location	Soil condition		P			S			E		
				Avg	Min	Max	Avg	Min	Max	Avg	Min	Max
Cabanbanan	N 14⁰ 14' 44.6"	Settled; land soaking	PMAM	27.7	7.3	49.4	19.2	-11.1	67.7	4.0	2.4	5.1
	E 121⁰ 26' 11.7"		AMPM	29.4	6.9	50.9	18.9	-20.3	78.6	4.3	3.3	5.9
Calachuchi	N 14⁰ 15' 35.8"	Settled; land soaking	PMAM	-0.6	-13.7	17.0	0.2	-16.3	17.3	2.6	0.3	4.6
	E 121⁰ 26' 42.9"		AMPM	-1.0	-13.9	18.9	0.6	-22.0	18.3	3.1	0.0	5.2
Biñan	N 14⁰ 15' 57.6"	Fallow; soaked for ploughing;	PMAM	4.8	-0.9	10.7	-3.4	-8.8	1.5	2.9	1.2	4.8
	E 121⁰ 26' 10.1"	saturated adjacent fields	AMPM	4.5	-9.6	16.6	-3.6	-12.0	1.2	3.2	1.4	7.9
Pagsawitan	N 14⁰ 16' 55.9"	Fallow; soaked for ploughing	PMAM	0.6	-1.3	3.8	6.7	0.1	14.7	3.0	1.5	4.7
	E 121⁰ 25' 32"		AMPM	0.7	-0.9	5.1	7.1	-2.8	29.3	2.7	0.0	4.6
Mayapa	N 14⁰ 16' 41.5"	Fallow; soaked for ploughing;	PMAM	1.3	-3.4	17.8	3.8	-14.0	33.7	3.6	0.6	6.8
	E 121⁰ 27' 07.3"	saturated adjacent fields	AMPM	1.1	-5.5	14.1	2.9	-12.9	25.1	3.3	1.1	5.4

Table 8.2. Results of percolation tests (mm day⁻¹) in Sta. Maria RIS

Site	Location	Soil condition		P Avg	P Min	P Max	S Avg	S Min	S Max	E Avg	E Min	E Max
Willingwiling	N 14° 26' 25"	Saturated;	PMAM	0.9	-1.2	4.7	-0.1	-1.5	1.3	2.2	1.1	3.1
	E 121° 25' 9.3"	newly transplanted	AMPM	0.8	-0.8	3.5	-0.1	-1.2	0.8	2.0	0.5	3.2
JP Rizal	N 14° 27' 55.8"	Saturated;	PMAM	0.3	-1.5	1.9	0.5	-0.6	3.3	3.1	0.8	6.4
	E 121° 25' 11.9"	newly transplanted	AMPM	0.3	-1.4	1.4	0.4	-0.6	2.0	2.7	0.2	5.1
Bagumbayan	N 14° 30' 20.1"	Fallow; soaked for	PMAM	0.3	-6.1	5.3	7.1	1.2	22.0	3.3	0.2	6.2
	E 121° 26' 20.4"	Ploughing	AMPM	0.4	-4.1	4.4	6.2	-2.6	28.4	4.6	5.9	2.7
		Saturated; vegetative	PMAM	1.3	-2.4	6.3	--	--	--	2.5	0.6	7.0
			AMPM	1.6	-0.9	4.7	--	--	--	2.7	0.8	7.2
Bubukal	N 14° 28' 41.6"	Fallow; saturated	PMAM	5.6	-0.2	12.5	-4.0	-9.5	5.4	2.6	0.3	6.6
	E 121° 24' 53.1"		AMPM	6.4	-0.2	9.4	-3.4	-7.9	1.4	2.5	6.6	0.0
Talangka	N 14° 28' 23.4"	Saturated; newly	PMAM	1.8	0.7	4.3	2.0	-0.4	6.1	3.5	2.2	5.9
	E 121° 26' 5.6"	Transplanted	AMPM	1.7	0.1	2.8	2.0	-0.9	7.5	3.4	2.2	6.2
Inayapan	N 14° 29' 35.0"	Fallow; soaked for	PMAM	4.1	2.2	7.3	5.8	2.5	9.3	3.0	1.6	4.8
	E 121° 24' 47"	Ploughing	AMPM	4.4	6.9	2.0	6.0	2.1	12.0	3.0	1.3	5.0
Bucol	N 14° 30' 2.9"	Settled, plowed	PMAM	2.8	-1.9	12.6	11.0	3.2	42.2	3.7	1.6	6.3
	E 121° 25' 2.3"	soil; land soaking	AMPM	2.9	-1.6	9.2	10.3	1.8	46.8	3.5	2.4	4.8
Coralan	N 14° 29' 0.2"	Settled, plowed	PMAM	3.9	0.4	7.2	15.1	3.0	36.9	3.2	1.0	6.2
	E 121° 24' 26"	soil; land soaking	AMPM	1.9	0.1	3.1	2.1	0.2	4.2	2.7	1.4	5.6
Johnson	N 14° 26' 58.5"	Fallow; soaked for	PMAM	3.3	0.6	5.7	2.3	-1.2	4.7	2.6	1.2	4.1
	E 121 25□0.3"	Ploughing	AMPM	3.3	0.6	5.4	2.6	-0.3	7.1	2.8	5.3	1.6

In Sta. Maria RIS, the measured average percolation rates in the 10 sites ranged from 1-6 mm day^{-1} (Table 8.2 and Figure 8.7). Only three of the 10 experiment sites (Bagumbayan, Wilingwiling and JP Rizal) had average percolation rates within the higher assumed value for the system of 1.4 mm day^{-1}. The highest average percolation value of about 6 mm day^{-1} was estimated for Bubukal site. Four sites (Inayapan, Coralan, Johnson and Bucol) had average percolation rates of 3-4 mm day^{-1}.

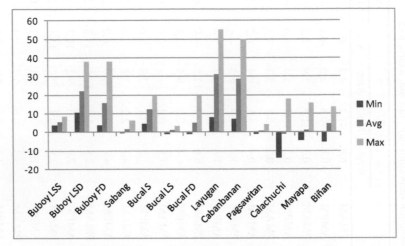

LS - Land soaking; LSS - Land soaking, settled soil; LSD Land soaking, disturbed soil; SS - saturated soil; FD - full duration

Figure 8.6. Ranges and averages of percolation rates in Balanac RIS experiment sites

The results of soil texture analyses (Tables 8.3 and 8.4) showed some differences between the soil samples from the experiment sites and their assigned soil textures as indicated on the official soil classification map of the Bureau of Soil and Water Management (BSWM). The most notable was the occurrence of sandy soils in Cabanbanan, Layugan and Biñan sites. Sandy soils were not among the reported soil textures, which range from clay to clay loam in Balanac RIS and clay in Sta. Maria RIS. The presence of sandy soils could explain the high percolation values obtained in Layugan and Cabanbanan. Sandy or light textured soils are associated with higher percolation and seepage rates as they offer less resistance to movement of water. Among the three sites with light-textured soils, Biñan had the lowest percolation values. The reason for this could be the continuous deep seepage flow and lesser resistance of soil below 30 cm depth (loamy sand) to such flow or even upswelling. The Biñan site was located downstream.

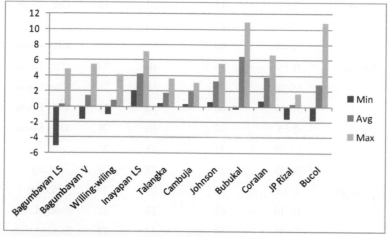

LS - Land soaking; V - vegetative growth

Figure 8.7. Ranges and averages of percolation rates in Sta. Maria RIS experiment sites

For sites with predominantly clayey soils, only Wilingwiling and JP Rizal had measured percolation rates within the assumed design value of 1.3 mm day^{-1} for clay soils. Meanwhile, majority of the sites with clay loam soils had percolation rates almost twice as much as the 1.8 mm day^{-1} design values for percolation.

The percolation rates obtained in the sites with sandy soils had 25-30 mm day^{-1}, which are much higher than the 2 mm day^{-1} conventionally used in the design for sandy clay loam soils and 4 mm day^{-1} for sandy loam soils. Aside from soil texture, other factors that could have influenced the percolation rates were the relative elevation of experiment site to the adjacent fields, welling up of shallow groundwater and ploughing that loosens or breaks the hard pan.

Table 8.3. Results of textural analysis for Balanac RIS experiment sites

Site		Percent Composition			Soil texture
		Sand (%)	Clay (%)	Silt (%)	
Bucal	0-15	41	29	30	Clay Loam
	15-30	44	26	30	Loam
	30-45	40	30	30	Clay Loam
Sabang	0-15	38	30	33	Clay Loam
	15-30	37	34	29	Clay Loam
	30-45	40	30	29	Clay Loam
Buboy	0-15	42	30	28	Clay Loam
	15-30	31	43	27	Clay
	30-45	19	57	24	Clay
Layugan	0-15	47	23	30	Loam
	15-30	61	10	28	Sandy Loam
	30-45	54	19	27	Sandy Loam
Cabanbanan	0-15	52	17	31	Sandy Loam
	15-30	57	15	28	Sandy Loam
	30-45	31	43	27	Clay
Calachuchi	0-15	31	38	32	Clay Loam
	15-30	28	42	30	Clay
	30-45	34	37	29	Clay Loam
Biñan	0-15	66	13	21	Sandy Loam
	15-30	74	10	17	Sandy Loam
	30-45	84	2	13	Loamy Sand
Pagsawitan	0-15	33	34	32	Clay Loam
	15-30	30	35	36	Clay Loam
	30-45	37	29	34	Clay Loam
Mayapa	0-15	34	39	27	Clay Loam
	15-30	37	40	23	Clay
	30-45	40	35	25	Clay Loam

Source: Abarabar (2014)

Table 8.4. Results of textural analysis for Sta. Maria RIS experiment sites

Site		Percent Composition			Soil texture
		Sand (%)	Clay (%)	Silt (%)	
Bagumbayan	10-20	40	35	25	Loam
	30-40	46	29	25	clay loam
Bucol	10-20	42	32	26	Loam
	30-40	41	27	32	clay loam
Inayapan	10-20	42	32	26	clay loam
	30-40	44	29	28	Clay
Coralan	10-20	39	38	23	clay loam
	30-40	40	31	29	Clay
Bubukal	10-20	32	34	34	clay loam
	30-40	42	34	24	Clay
Talangka	10-20	42	35	23	Clay
	30-40	42	32	27	Clay
JP Rizal	10-20	37	32	32	Clay
	30-40	41	27	32	Clay
Cambuja	10-20	37	32	32	clay loam
	30-40	49	29	22	clay loam
Johnson	10-20	38	35	26	Clay
	30-40	45	27	28	Clay
Wilingwiling	10-20	42	31	26	Clay
	30-40	47	37	16	Clay

Source: Avila (2014)

8.3.2   *Results of the ponding tests*

The results of the ponding experiments showed a wide range of magnitudes of conveyance losses in the representative farm ditches of Balanac RIS and Sta. Maria RIS (Tables 8.5 and 8.6). Considering the measured percolation rates for each experiment site, most of the conveyance losses can be attributed to seepage. Most of the measured farm ditch losses were not necessarily lost to the systems as most of the canal water flowed into adjacent paddy fields. Nevertheless, considerable seepage must be managed as it affects timely delivery of water to downstream farms. Unlike the rest of the ponding sites, the ponded farm ditch in Layugan runs along the border of the irrigation service area, adjacent to an orchard, an unsaturated field. How such situation was taken into account in the estimation of design water duty is worth reconsidering.

Table 8.5. Farm ditch conveyance losses in m^3/day/km based on daytime and 24-hr continuous measurements in Balanac RIS

	Daytime			Daytime + night time		
	Ave	Min	Max	Ave	Min	Max
Bucal[1a]	398	344	483	--	--	--
Sabang	76	38	91	47	32	53
Buboy	64	27	101	57	31	81
Layugan[1a]	190	64	574	--	--	--
Cabanbanan[1a]	99	49	173	--	--	--
Calachuchi	26	-51	193	12	-9	52
Biñan	16	-7	60	8	-7	27
Pagsawitan[1b]	116	95	137	--	--	--
Mayapa	13	-12	43	11	0	31

[1]24-hr continuous measurement was impractical because the ponded water practically drained from the canal within few hours due to (a) high percolation and seepage and (b) paddy eels inhabiting the canals and creating burrows during night time.

8.4 Estimates of crop water requirements and irrigation requirements

As part of the operation, maintenance and management plan for the irrigation systems, the concerned senior water resource facilities technicians (SWRFT) and senior engineers of the system operation division of the Irrigation Management Office calculate the extent of the service area that can be irrigated for the forthcoming wet and dry cropping seasons. The extent of the area that can be served, also called the programmed area, for each cropping season is determined based on the assumed available water for diversion, soil saturation requirement, paddy field submergence, percolation, evaporation or evapotranspiration, farm waste and distribution losses, conveyance losses and effective rainfall. It is the total of the estimated weekly areas that can be land soaked during the 3-week and 10-week land soaking periods in Balanac RIS and Sta. Maria RIS, respectively.

Table 8.6. Farm ditch conveyance losses in m^3/day/km based on daytime and 24-hr continuous measurements in Sta. Maria RIS

	Daytime			Daytime + night time		
	Ave	Min	Max	Ave	Min	Max
Bagumbayan	8	-15	36	3	0	10
Bubukal	132	49	186	43	16	60
Bucol	37	2	134	12	3	33
Inayapan	145	86	209	59	29	100
Coralan	102	13	280	34	14	50
Cambuja	68	1	182	24	2	40
Johnson	65	13	152	20	12	26
JPRizal	190	42	419	65	33	96

The crop water requirement (CWR) is computed by the concerned system personnel as the algebraic sum of evapotranspiration and percolation. Constant crop coefficient values of 1.25 for both cropping seasons for Balanac RIS and 2.1 and 1.6 for wet season and dry season cropping, respectively, for Sta. Maria RIS were used (Table 8.7). The values of other water balance parameters used in the calculations of CWR and irrigation water requirements are lifted from old project documents and records for the systems. These values are used as "standard" values and assumed constant throughout the given cropping season.

The significantly higher observed percolation rates in 5 of the 9 experiment sites in Balanac RIS and in 5 of the 10 sites in Sta. Maria RIS (Figures 8.8 and 8.9) would translate to higher CWR than that used during the planning stage. With the observed average percolation rates of 2-15 and 2-5 times higher than the design values used for Balanac RIS and Sta. Maria RIS, respectively, under estimation of CWR was more serious in the former.

The observed farm ditch conveyance losses were way greater than the assumed values during the planning stage. The validity of the generally assumed 70% application efficiency or, in this case, Farm waste + distribution losses (FW+DL) equal to 30% of the crop water requirements, has been questioned by many experts (Bos and Nugteren, 1990; FAO, 1992; Plusquellec, 2002). While a significant portion of the measured farm ditch conveyance loss can be attributed to seepage into the adjacent paddy fields and were not physically lost from the system, such magnitude of farm ditch conveyance losses lead to underestimation of irrigation water requirements at and irrigation duration for the farm turnouts.

Table 8.7. Values of water balance parameters used for purposes of design and O&M planning for Balanac RIS and Sta. Maria RIS

Parameters	Balanac RIS		Sta. Maria RIS	
	Wet season	Dry season	Wet season	Dry season
Evaporation, mm day^{-1}	4	4	3.09	4.08
Evapotranspiration, mm day^{-1}	5	5	6.36	6.40
Percolation, mm day^{-1}	2	2	1.30	1.41
Crop water requirement (CWR), mm day^{-1}	7	7	7.66	7.81
Soil saturation requirement, mm	64.3	66	50	66
Submergence for rice cultivation, mm	50	50	50	50
Effective rainfall, mm day^{-1}	--	--	1.35[a]	3.17[a]
			1.20[b]	3.19[b]
Farm waste + distribution losses (FW+DL), % CWR	30	30	30	30
Conveyance loss, % CWR	20	20	18	20
Water supply, l s^{-1}	3,500	3,500	418[c]	437[c]
			389[d]	512[d]

[a]Wet season: May - October; [b]Dry season: November - April; [c]Bagumbayan; [d]Mata

8.5 Assessment of the reliability of the estimates on available water supply

Actual water diversions of Balanac RIS were not measured. Without gauging stations, system officials for Balanac RIS acknowledged that the values they were using for available water supply were basically a result of guess work based on the approved water right of the irrigation system. The adopted values of available water supply for Sta. Maria RIS were only about 40% of the permitted diversions. They were based on the results of one-time system initiated investigation of discharge-head relationships and observed usual levels of water flowing through the flumes immediately downstream of the main intakes of the two dams. The discharge rating tables determined for the flumes prior to 2000 were available at the NIA System Office. The records of water levels at the flume were being monitored by the irrigators association since 2002. However, there were non-ideal conditions for accurate flow measurement by using Parshall flumes seen during the field investigation. These included untrimmed vegetation along the approach canals, sedimentation obstructing the converging section, and rock-filled bottoms at throats and converging sections.

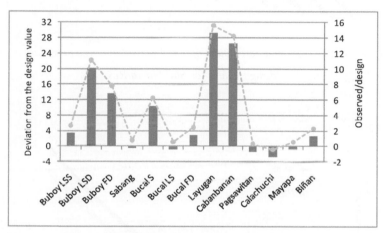

Figure 8.8. Deviations from and percentage of the observed average percolation rates to the design value for Balanac RIS

8.5.1   Discharge-head relations

The discharge-head relations were generated based on these discharge rating tables (Figure 8.10). The canal discharges as a function of head at the respective intakes of Sta. Maria Main Canal (SMMC) and Mata Main Canal (MMC) could be estimated by the equations $Q = 2530H^2 - 793H + 1.76$ and $Q = 2580H^2 - 459H + 1.64$, respectively. The IA-measured water levels at the flumes were plugged into these discharge-head equations to calculate the discharges, which were then used to generate the corresponding frequency distribution curves (Figure 8.11).

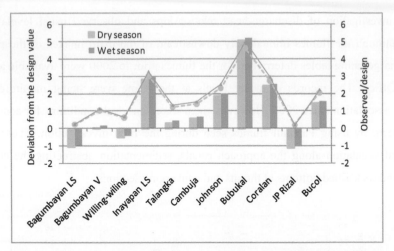

Figure 8.9. Deviations from and percentage of the observed average percolation rates to the design values for Sta. Maria RIS

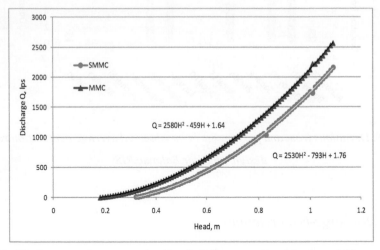

Figure 8.10. Discharge rating curves for Parshall flumes at Sta. Maria and Mata main canals intakes (Data source: NIA Sta. Maria RIS Field Office)

Based on the generated frequency curves, the adopted values for available water supply for SMMC and MMC have a probability of exceedance of about 33% and between 14-18%, respectively (Figure 8.11). The flow magnitudes associated with the conventional design probability of 80% were only about 200 l s^{-1} for SMMC and 60 l s^{-1} for MMC. However, such very low flow magnitudes and low probability of the observed "usual flows" deviated from the local knowledge on available water and from the fact that the irrigation system had consistently irrigated an area of around 850 ha.

8.5.2   Field measurements of diverted flows

In the absence of relevant flow records, hydrologists and engineers of the NIA take discharge measurements during dry months and consider the measured flow as the sustained flow or available water to an irrigation scheme. The size of the area that can be irrigated is estimated based on this sustained flow and irrigation water requirement (Personal communication). Dry season flow measurements were also carried out in Sta. Maria RIS to approximate its low flows. The discharges (Table 8.8) were estimated by using the area-volume method, which uses a current meter to measure flow velocity. The results of sample flow measurements showed that the magnitude of available water supply used in projecting the areas that can be irrigated by Santa Maria Main Canal during dry seasons were within the range of the measured flow rates. In contrast, the assumed available water for the Mata Main Canal during dry seasons was higher than most of the measured values.

8.5.3   Programmed areas versus actual irrigated areas

In general, a comparison between the projected area for irrigation and the actual area served would be an indication of the reliability of the assumed values of available water supply. In the case of Balanac RIS and Sta. Maria RIS, their respective programmed areas or target areas for irrigation were achieved better during the dry season than during the wet season (Figure 8.12). These trends were attributed to less certainty of the spatial extent of service areas that will be submerged due to heavy downpours and floods during the monsoon months.

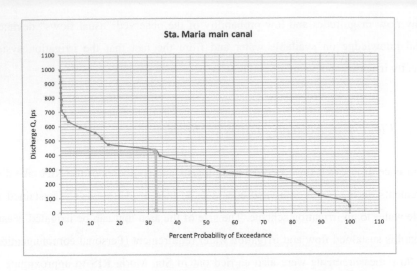

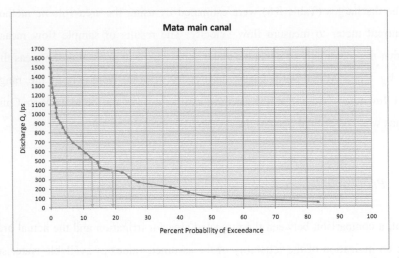

Figure 8.11. Flow duration curve for Sta. Maria Main Canal and Mata Main Canal intakes

Further, there were noticeable gaps between irrigated and benefitted areas, especially during the wet cropping seasons. Benefitted areas were irrigated areas with rice yield of above 2 tons/ha. Farmers with lower yields were exempted from paying irrigation service fees. The low yields in "irrigated-but-not-benefitted" areas were attributed to monsoon-related damages, pests, diseases, and water shortage (Figure 8.13 and Table 8.9). In both systems, it was mainly due to monsoon-related causes such as submergence and typhoon-lodged rice plants during the wet season. The main causes of damages in Balanac RIS and Sta. Maria RIS during the dry season were water shortage and pests and diseases, respectively.

Table 8.8. Sample measurements of diverted flows in Sta. Maria main sanals

Dry season cropping	SMMC	MMC
28-Nov-12	796	
30-Nov-12	895	482
03-Dec-12	847	
04-Dec-12	668	
11-Dec-12	613	528
19-Dec-12	568	496
16-Jan-13	577	884
15-Mar-13	164	320
21-Mar-13	563	263
06-Apr-13	552	307
18-Apr-13	404	
23-Apr-13		260
05-Mar-14	366	
13-Mar-14		325
14-Mar-14	410	
21-Mar-14		407
27-Mar-14	560	
28-Mar-14		509
08-Apr-14	271	
23-Apr-14	220	
28-Apr-14		297
Wet season cropping		
07-May-14	168	
22-May-14		230

Unit: $1 s^{-1}$

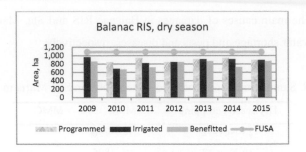

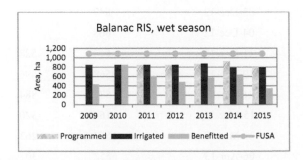

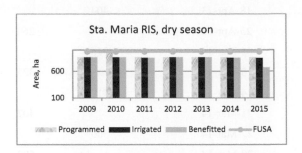

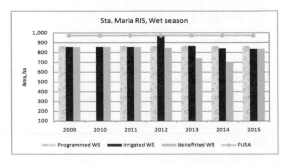

Figure 8.12. Programmed, irrigated, benefitted areas and FUSA

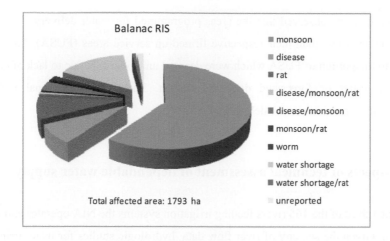

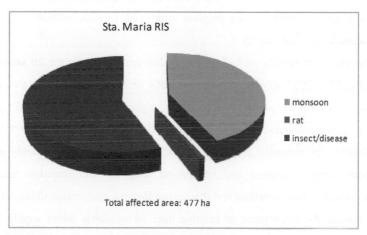

Figure 8.13. Affected areas and causes of crop damages in Balanac RIS and Sta. Maria RIS during 2008-2015

It is interesting to note that while it is a common knowledge among farmers and system operators that Balanac RIS had more available water supply per ha than Sta. Maria RIS, exemption from irrigation service fee (ISF) payment due water shortage was reported in the former. Sta. Maria RIS practiced segmental relay planting and rotational irrigation to circumvent their limited water supply. Such cropping practice, which provided continuous presence of host plants to pests and diseases, was viewed to be the reason for the major cause of crop damages.

Also, it can be observed that the areas programmed for water delivery service in both systems remained less than their respective firmed-up service areas (FUSA). The difference pertained to the portion of FUSA which were deemed unserviceable due to lack of operational irrigation facilities or forecasted drier dry season. These non-operational FUSA were excluded in the irrigation service delivery schedule.

8.6 Prospects of technical assessment of dependable water supply

Only about a third of the 165 rivers feeding irrigation systems the NIA operates and maintains are gauged. Given the scarcity of river flow data, hydrologic studies for many irrigation and multipurpose irrigation infrastructure projects employed various methods or techniques of estimating dependable water supply.

The results of the review on hydrologic studies carried out in the 20 sample irrigation projects during 2010-2015 showed that the methods used included one or a combination of the following: empirical and statistical techniques, regionalization, conceptual rainfall-runoff relations, spot flow measurements and local knowledge of the seasonal flows to assess water supply availability for irrigation projects (Annex G). The methods commonly utilized were regionalization using regression and correlation analyses and drainage area ratio and proportion to runoff, rational method and dry season flow measurements using current meters.

Cognizant of the importance of reliable data on available water supply to irrigation system modernisation plans, the potentials for technical prediction of flows of Balanac River and San Antonio River, which feed Balanac RIS and Sta. Maria RIS, respectively, were investigated. The results of a survey of official and popular data sources on the availability of commonly required information for estimating river flows for ungauged rivers indicated that flow estimates obtained by using straight forward application of the commonly used regionalization method would be fraught with considerable uncertainty (Table 8.9). Issues on the accuracy of the estimates would stem from the scarcity of relevant data inputs for flow assessments. An extrapolation of flows for recent decades based on empirical and statistical relations of recorded data prior to 1980s imply the assumption of stationarity of river basins behaviour and response to factors affecting runoff. The local knowledge on the significant changes in land use in the river basin, climate and river flow regimes refutes such assumption.

Table 8.9. List of available official information for dependable flow estimation for Balanac
RIS and Sta. Maria RIS

	Format/period	Resolution	Sources
Climatic data			
• Rainfall, mm	1917-1938	Monthly	NAS-UPLB
	1947-present	Daily, monthly	
• Evaporation USWB, mm	1959-present	Daily, monthly	NAS-UPLB
• Relative humidity, %	1954-present	Daily, monthly	NAS-UPLB
• Temperature, °C	1954-present	Daily, monthly mean, maximum and minimum	NAS-UPLB
River basin			
• Topography	Paper map, JPEG; 1947-1953, 1979, 1990, 1992	1:50,000; 20 m contour interval, 5-10 m supplementary contour	NAMRIA
	IFSAR DTM	5 m	NAMRIA
• Soil map, soil type	Paper map, TIFF; 1946	1:100,000; by province	BSWM
• Slope map	Paper map, TIFF; 1955-1956, 1970	1: 50,000	BSWM
Streamflow:			
• Balanac River	m^3s^{-1}, 1958-1980	mean monthly, annual	NWRB, BRS
• Mayor River	m^3s^{-1}, 1958-1970; 1983-1993; 1994-2007	mean monthly, annual	NWRB, BRS

NWRB - National Water Resources Board; BRS - Bureau of Research and Standards; DPWH - Department of Public Works and Highways; PAGASA - Philippine Atmospheric, Geophysical and Astronomical Services Administration; NAMRIA - National Mapping and Resource Information Authority; BSWM - Bureau of Soil and Water Management; IFSAR-DTM - Interferometric Synthetic Aperture Radar - Digital Terrain Model

Nevertheless, applicability or apppropriateness of available flow estimation techniques and simulation tools with the given sparce river flow and other data inputs must be explored. River flow gauging would need to be carried out to support validation of the flow estimates.

It would need to be continuing activity in order to start building a hydrologic database for more reliable flow estimates. Information on the available streaflow records for rivers within the river basin or nearby similar basins would be relevant in designing for the methodology and selecting for tools for available irrigation water supply assessment study. Such information was gathered from the DPWH and NWRC and assembled for the case study systems (Annex H).

8.7 Conclusion

The results of in-situ measurements of percolation revealed different rates in different parts of the service areas of Balanac RIS and Sta. Maria RIS. The deviations of measured percolations from the assumed values of 2 mm day^{-1} for Balanac RIS were higher by 5-14 times of the former in four of the experiment sites. In Sta. Maria RIS, the observed deviations in four of the measurement sites were higher by about 2-5 times of the assumed rate of 1.3-1.4 mm day^{-1}. The significantly higher actual percolation rates will result in gross underestimation of crop water requirement and irrigation requirement. The wide range of measured percolation rates within the service areas, especially in Balanac RIS, points to the deficiency of a design and planning approach of assigning a percolation value for the whole service area based on an assumed homogeneity or generalized characterization of the soil in the area. The big difference between the assumed and actual percolation rates presents a strong case for the need to determine the actual value in at least one representative site per turnout service area.

The magnitudes of diverted flows computed by using the discharge-head relations for Sta. Maria RIS main canals and the corresponding water level data were much lower than the available flows used in projecting the area to be irrigated. The assumed available flows for Sta. Maria and Mata main canals have probabilities of exceedance of only about 33 and 16%, respectively. Such low probabilities ran counter to the local knowledge and general field observations of the "normal" flows along the main canals. The low probabilities of the assumed flow magnitudes that resulted in dry season actual irrigated areas within the neighbourhood of the figures for dry season programmed areas would mean that the discharge-head relations were no longer valid. The results of sample flow measurements lent credence to the validity of the assumed flows based on field experience and local knowledge on the usual flow magnitudes of SMMC. However, the results of similar flow measurements did not support the assumption on the water availability in the case of MMC.

Reliable data on available water for an irrigation system are crucial inputs to the formulation of options for system modernisation, especially those aimed at expanding the service area and increasing cropping intensity. Availability of water would be a rational criterion in prioritizing rehabilitation and modernisation project proposals for budget allocation. Results of preliminary investigation showed that Balanac RIS has relatively good potentials to adequately meet its present irrigation demand as well as that of possible service area expansion. In contrast, expanding the irrigation coverage of Sta. Maria RIS was constrained by limited water supply at the source. To better quantify the available water and fine-tune modernisation plans, in-depth technical assessments of discharges of Balanac River and Sta. Maria River must be pursued. The utility of the regionalization method by using similar river basins belonging to same climate type in adjacent regions need to be investigated. Likewise, the use of geographic information systems (GIS) and remote sensing data, which have been successfully used in other countries in quantifying the required river basin parameters, would have to be explored. More importantly, a river flow gauging program needs to be carried out.

9 CHARACTERIZATION OF THE SYSTEM MANAGEMENT, SERVICES AND DEMAND TO IDENTIFY MODERNISATION POTENTIALS

9.1 Introduction

Comprehensive knowledge about the water flow paths, O&M, irrigation service, unit service areas and demand for operation is a crucial input to the formulation of appropriate and workable plans for irrigation system modernisation (Renault *et al.*, 2007). Good understanding of the existing and potential sources of inflows and outflows in the service area, their routes, points of convergence, flow magnitudes and timings is specifically important in developing water distribution and allocation strategies, evacuation of flood water and identifying locations for water reuse facilities. Meanwhile, knowing the current resource inputs and cost of O&M lends insights on the cost-effectiveness of current operation and maintenance and consequently, on identifying cost-saving items and technical options for a more cost-effective operation. Also, it provides information on setting a deliverable service level and price of water delivery service. On the other hand, awareness of the quality of current irrigation service and types of services to other users within the service area, or even at the basin level will facilitate a more holistic examination of the present state of service as well as the challenges and prospects for improving and expanding it.

One kind and level of irrigation service requirement is easier to fulfill than multiple demands. Thus, from a management perspective, partitioning of the service area into sub-areas based on homogeneity of desired service, hydrologic characteristics and level of participatory management is desirable as it would allow management of each partitioned area as a single management unit. This would promote greater responsiveness of operation to the unit-specific irrigation service needs and a more efficient system and water management. Moreover, assessment of the corresponding operational needs to achieve the set targets of water delivery service at key distribution points is critical to rational allocation of irrigation management resources such as manpower, skills, transportation, communication, water distribution structures and monitoring equipments, among others.

The MASSCOTE approach involves spatial analyses of the information on water network, management unit, O&M, irrigation service and demand, among others, in crafting for a modernisation plan for an irrigation system. This chapter is aimed at applying its analytical concepts and tools while crafting appropriate adaptive-modifications to suit the case of small-scale national irrigation systems in the Philippines.

9.2 General research methodology

Characterization of the water network, system operation, maintenance and management and irrigation service in the context of identifying potential improvements of present and possible future scenarios necessitated analyses of considerable system-specific data and information. However, relevant data and information are scarce and not systematically organized in the case of Balanac RIS and Sta. Maria RIS. This may be attributed to the fact that system management is more focused on day-to-day field operation and institutional administration. Data collection, collation and generation were carried out in preparation for irrigation system characterization and analysis.

Composite maps of water flow path networks were GIS-generated based on a 5 m resolution digital elevation model (DTM) sourced from the National Mapping and Resource Information Authority (NAMRIA), Google earth images of the irrigation service areas, paper maps and AutoCad files of the general irrigation system layout. GIS-usable files of the relative locations of current tertiary service areas, or management units within the service area were generated based on the system paper parcellary maps, global positioning system (GPS) data and Google Earth delineation technique.

The water sources and immediate supply canals, percentage of area supplied by each source and seasonal water supply situation were identified for each TSA. Well drillings were carried out in two sites in Sta. Maria to investigate the potentials of its groundwater resource for supplemental irrigation and to identify possible schemes for conjunctive use of pumped water and dam-diverted water in improving water delivery service.

The cost of management, operation and maintenance were determined based on the financial statements of the water users associations and the NIA-system related expenses. Percentages of expenses for the different cost items were determined. Actual expenses were compared with country average and standard values adopted by NIA for planning purposes.

The relative quality of irrigation service was assessed in terms of adequacy, reliability, equity, flexibility and measurements of water delivery (Annex I). The assessment made use of the system data gathered through questionnaires, system walkthroughs and field interviews of system operators and turnout service area (TSA) leaders of the water users associations. Except for equity, each of the indicators of quality of irrigation service was rated according to the rapid appraisal procedure (RAP) rating scale of 0 to 4 (with 0 the worst and 4 the best) for actual water delivery service. The equity was defined based on the sufficiency of the irrigation supply available to a turnout service area vis-a-vis that of most upstream areas. It was rated in a scale of 0 (very low compared to upstream) to 4 (same as that of upstream farms) by the farmer-leaders of each turnout service area (TSA). The overall values indicating the quality of service for each TSA were computed following the RAP method. Spatial distributions of the quality of water delivery service as well as those of other categories of water uses within the service areas were identified.

The demand for canal operation was qualitatively assessed based on the three factors proposed in MASSCOTE - demand for the service, perturbation and sensitivity (Annex I). The demand for service was determined based on the level of crop risk or vulnerability due to water scarcity, absence of other sources of supplemental water and occurrence of flood. A value of 1 was assigned to each risk situation existing in the TSA of interest. Due to the lack of relevant flow records and infeasibility of flow measurements to determine sensitivity and perturbations, indicators were used instead. Diversion sensitivity or hydraulic flexibility was determined based on the hydraulic type and operational status of the gates while level of perturbation was based on the number of ungated offtakes and inlets upstream of the gate or structure of interest. A rating scale of 1 (low) to 3 (high) was used for both. The aggregated values of the three factors of demand for canal operation were reclassified into 5 groups indicating the different levels of demand for canal operation.

9.2.1   Water network

The geographic locations of the irrigation network and developed drains are quite well known to systems operators and farmers. However, this is not the case for natural drains or waterways within the service areas. The composite map of canals and drainage networks, including natural drains and waterways and contour maps of Balanac RIS and Sta. Maria RIS lend information where water is drained and, relatively, how much drainage water flows at specific locations (Figures 9.1 and 9.2). The use of 5 m resolution DEM for irrigation service areas would greatly facilitate the task of identifying potential sites for water reuse facilities as compared to the use of conventional paper maps generated during the project planning phase and feasibility study for the original system (Annex J). Morever, such old maps may no longer exist at NIA offices as in the case of Balanace RIS.

9.2.2   Cost of management, operation and maintenance

The expenses incurred in providing water delivery service typcially include salaries, wages, insurance, repairs and maintenance, travels and office supplies, among others. In the case of the Laguna-Rizal Irrigation Management Office (IMO), which manages Balanac RIS and Sta. Maria RIS among other systems, cost accounting of irrigation management and operation per system is not done since basically its same set of personnel, equipments and other resources are used for managing all irrigation systems under its jurisdiction. In short, the IMO is a common resource pool for the irrigation systems under its jurisdiction.

To determine the cost of operation and maintenance (O&M) of the case study systems, a distinction has been made between expenses incurred in providing frontline water delivery service or basic system operation and general management or administration. The former was defined as the system O&M cost. In the case of Balanac RIS and Sta. Maria RIS, which are under Irrigation Management Transfer (IMT) contracts, the O&M cost was grouped into two: the NIA-based expenses and the water users' association (WUA) expenses (Table 9.1). The former mostly included expenses for NIA personnel directly involved in the system operations. The latter included the cost of water delivery service and system repairs and maintenance (details in Annex K).

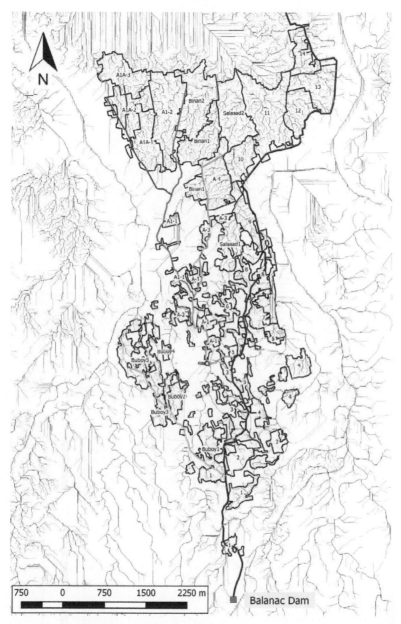

Figure 9.1. The drainage network in the different turnout service areas of Balanac RIS

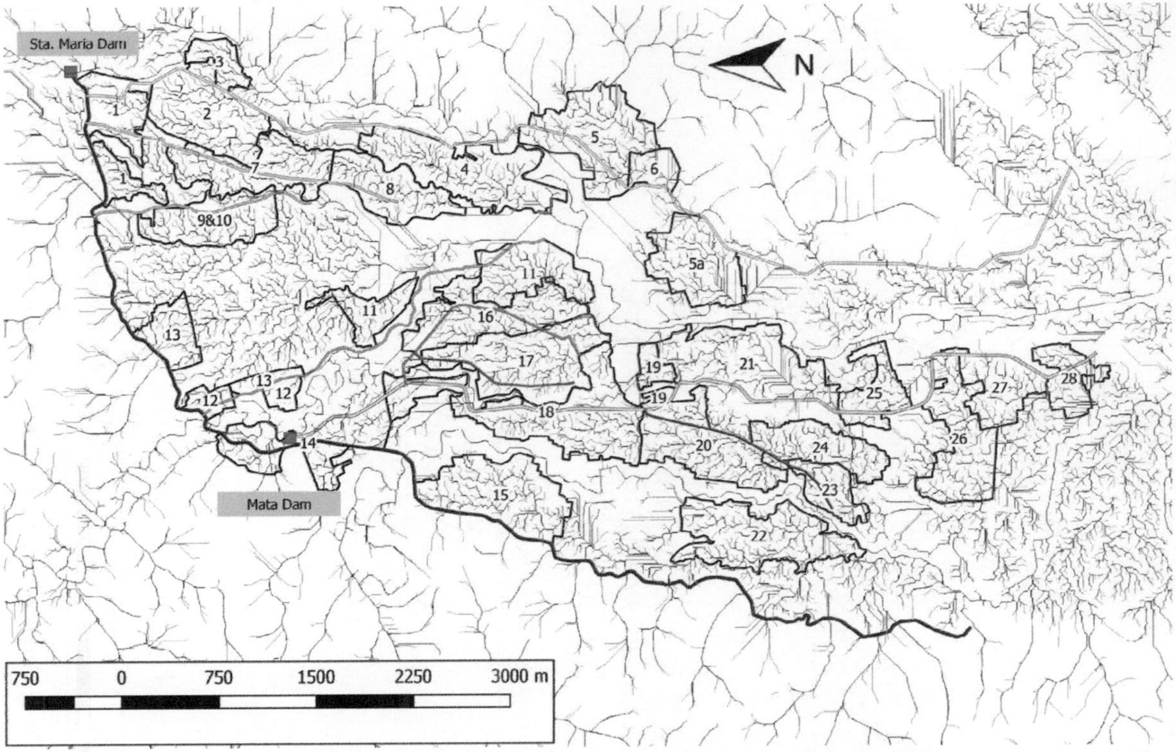

Figure 9.2. The drainage network in the different turnout service areas of Sta. Maria RIS

Table 9.1. Average O&M expenses for Balanac RIS and Sta. Maria RIS

Cost items	Balanac RIS (2012-2015)		Sta. Maria RIS (2010-2015)	
	NIA	BRISIA	NIA	SANTAMASI
Personnel	451,781	574,604	231,069	531,530
Repairs & maintenance	-	85,035	-	132,911
Office administration	-	55,256	-	35,356
Travel	-	66,918	-	101,629
Miscellaneous	-	120,721	-	28,058
Lumped non-personnel[1]	161,029	-	35,094	-
Total	612,810	902,534	266,163	829,484

Source: Laguna-Rizal IMO Accounting Office files and WUA office files for different years
[1]Prorated based on the system FUSA

The average annual cost of frontline O&M for Balanac RIS was about 40% higher than that for Sta. Maria RIS for their respective period-of-analysis. The O&M costs per ha for Balanac RIS and Sta. Maria RIS were about PhP 1,500 and PhP 1,100 (Figure 9.3). These expenses were lower than the standard cost per ha (PhP 2,000-3,000) for O&M used by NIA for planning purposes. Also, these costs were lower than the PhP 2,300 ha^{-1} average national O&M budget and the PhP 1,900 ha^{-1} average national actual expenses (Figure 9.4).

The personnel-related expenses accounted for about 70% of the O&M expenses of each system. The slightly higher personnel expenses for Sta. Maria RIS were due to the incentives its WUA gave to its officers and workers who met their O&M targets. In contrast, Balanac did not have an incentive system for its personnel. About 12% of the budget of Sta. Maria RIS was spent for repairs and maintenance of its canal network, while only about 6% was spent in Balanac RIS. The higher travel expenses of Sta. Maria RIS were mainly due to the fuel allowance given to its canal operators or water distribution men-cum-ISF collectors. There were no fuel allowances given for their couterparts in Balanac RIS.

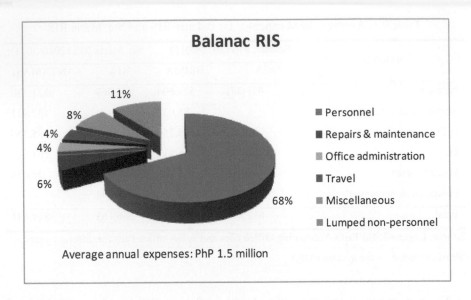

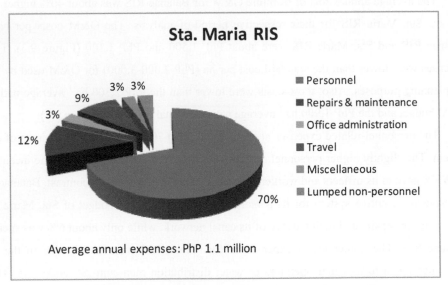

Figure 9.3. Percent distribution of average annual expenses for Balanac RIS and Sta. Maria RIS for the period 2012-2015 and 2010-2015, respectively

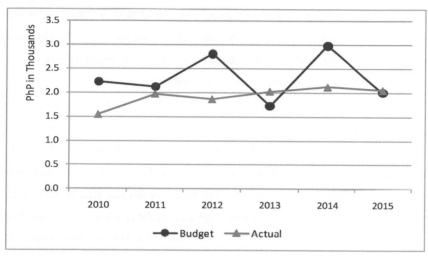

Figure 9.4. National average O&M budget and actual expenses per ha (Data source: NIA System Management Division (SMD) Performance Report, various years)

The average miscellaneous expenditure of Balanac RIS was at least four times higher than that of Sta. Maria RIS and was the second highest cost item for the system. Also, the NIA spent at least four times more for its O&M than that for Sta. Maria RIS.

While the relatively low O&M budget for both systems present a basis for an assumption that the budget was spent on the basic necessities for O&M, there were cost items that would need to be examined when identifying a possible cost-effective O&M plan. Miscellaneous expenditure in Balanac RIS was twice the cost incurred for a more legitimate and defined canal repairs and maintenance expenditure. The average expenditures for personnel services of almost 70% in both systems were significantly higher than the prescribed and agreed percentage allocation for such expenses, that is, 40% of the WUA budget. This is so even after excluding the costs of incentives and training and despite the fact that only NIA field personnel who are directly involved with the system operations and only IMT-prescribed WUA personnel composition were considered in the computation of personnel expenses.

The issue of inadequate O&M budget of publicly funded irrigation systems has been raised by a number of researchers and irrigation professionals (McLoughin, 1988; Wijayaratna and Vermillion, 1994; Vermillion, 2005; Burton, 2010; Malik *et al.*, 2014).

Questions on how much is a reasonably adequate budget for O&M activities that would result in the achievement of the target irrigation system performance has been asked by policy- and decision-makers. However, such topic is outside the scope of this research study.

Under the NIA IMT arrangement, the WUA budget mainly comes from its share of the ISF collection. The sharing ratio between the NIA and WUA varies depending on the IMT stage level, hence, delineation of respective O&M responsibilities. In case of Balanac RIS and Sta. Maria RIS, the ISF shares of the WUA were in effect the frontline O&M budget for the systems as their respective WUA spent it to carry out canal operations, water distribution and ISF collection functions. For 2010-2015, the ISF shares of Balanac RIS WUA averaged PhP 1.07 million or about 53% of collected ISF (PhP 2.02 million) while that of Sta. Maria RIS was about PhP 839,000 million or 38% of the PhP 2.20 million ISF collection (Figure 9.5).

The total ISF payment or collection in Balanac RIS for the period 2005-2015 averaged about 64% of the total collectible ISF. It was about 59% of the payable or collectible ISF in Sta. Maria RIS. During the same period, the amount of ISF collected natiowide averaged about 61% of the total collectible ISF (Figure 9.6).

In general, an increase in ISF collection means an increase in ISF share, hence, in WUA budget for O&M. It is possible with the increase in collectible ISF, which is dependent on the increase in benefitted areas or irrigated areas that attained a rice yield of more than 2 ton ha^{-1}. During the dry season, the amount of collectible ISF is proportional to the areas irrigated. This is not the case during the wet season as moonson-related production constraints such as typhoons, floodings, pests and diseases result in low yield or no harvest in some parts of the service area.

A shift or adjustment in cropping calendar to avoid destructive periods of the moonson season would help to minimize cases of poor harvest, thus, exemption from ISF payment. Its possibility has to be studied. Further, a strategy or a program aimed at encouraging ISF payment and collection has to be part of a comprehensive modernisation plan.

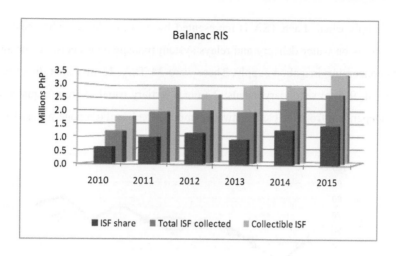

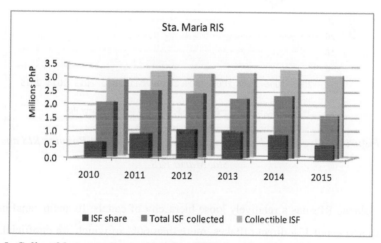

Figure 9.5. Collectible irrigation service fees (ISF) based on benefitted area, collected ISF and ISF shares of Balanac RIS and Sta. Maria RIS

9.3 The management unit and water distribution

9.3.1 Balanac RIS

The service area of Balanac RIS is partitioned into turnout service areas (TSA) based on hydraulic boundaries (Figure 9.7). The TSA are numbered based on their respective immediate supply canal. Each TSA is represented by a farmer-leader who speaks for TSA farmers concerns on water delivery and relays system management decisions, or agreements on matters concerning irrigation supply. There were 34 TSA, 15 of which divert water from the main canal; 9 from secondary canals; 2 from a tertiary canal; 3 from a quaternary canal and 5 are independent from the canal network, tapping creek and drainage water (Table 9.2).

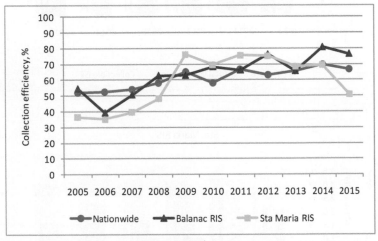

Figure 9.6 Average ISF payment at country level and in Balanac RIS and Sta. Maria RIS

The Balanac RIS has a relatively loose hierarchy of canals. Its main canal has 18 main farm ditches and about 118 direct offtakes, more than 60% of which are unofficial. Similarly, the lower level conveyance canals have a relatively high proportion of direct outlets.

The operation of dam gates and care of water distribution along the main conveyance canals were responsibilities of a NIA gate keeper and the water users association of Balanac RIS (BRISIA), respectively. The gates of the main intakes are open most of the time and closed only during occurrence of floods spewed by typhoons and heavy monsoon rains. In general, the wet cropping and dry cropping seasons are from May to October and November to April, respectively.

Laissez faire planting calendar is practiced by farmers. There is no water distribution plan or agreed water delivery schedule prior to the start of the cropping seasons. The actual water distribution is quasi continuous. It is continuous for the upstream reach of the main canal and either continuous but reduced flow or rotational for the downstream reach, depending on the water supply situation. Rotation and reduced, continuous flows are often practiced during the dry season and in cases of strong demand from TSA leaders or farmers. Two ditch tenders of BRISIA attend to water deliveries: one for the main canal and the other for lower level canals. They used makeshift checks at key bifurcation points to effect water distribution.

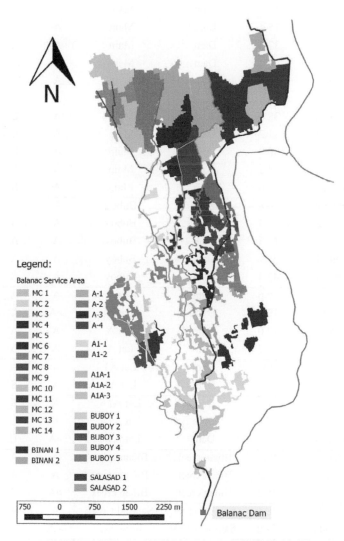

Figure 9.7. Turnout service areas of Balanac RIS

Table 9.2. Profiles of Balanac RIS tertiary service areas

TSA No	TSA, ha	Water source	Immediate supply canal	Water supply situation WS	Water supply situation DS
MC 1	33.0	Dam	Main	A	A
MC 2	20.0	Dam	Main	A	A
MC 3	17.9	Dam	Main	A	A
MC 4	32.7	Dam	Main	A	A
MC 5	15.7	Dam	Main	A	A
MC 6	15.4	Dam	Main	A	A
MC 7	37.6	Dam	Main	A	A
MC 8	28.4	Dam	Main	A	A
MC 9	23.4	Dam	Main	A	A
MC 10	30.5	Dam	Main	A	A
MC 11A	38.9	Dam	Main	A	L
MC 11B	31.2	Dam	Main	A	L
MC 12	30.0	Dam	Main	A	L
MC 13	44.2	Dam	Main	A	L
MC 14	18.6	Dam	Main	A	S
Buboy 1	20.8	Dam	Buboy	A	A
Buboy 2	13.3	Dam	Buboy	A	A
Buboy 3	12.3	Dam	Buboy	A	A
Buboy 4	9.6	Dam	Buboy	A	A
Buboy 5	43.4	Dam	Buboy	A	A
Lat A1	10.6	Dam	Lat A	A	A
Lat A2	10.8	Dam	Lat A	>A	L
Lat A3	12.0	Dam	Lat A	A	A
Lat A4	44.7	Dam	Lat A	A	A
Lat A1-1	4.9	Dam	Lat A1	A	A
Lat A1-2	74.1	Dam	Lat A1	A	S
Lat A1A-1	39.3	Dam	Lat A1A	A	S
Lat A1A-2	37.4	Dam	Lat A1A	A	L
Lat A1A-3	52.1	Dam	Lat A1A	>A	L
Biñan 1A	10.9	Biñan creek	Biñan creek	A	A
Biñan 1B	42.2	Biñan creek	Biñan creek	A	A
Biñan 2	75.2	Biñan creek	Biñan creek	>A	S
Salasad 1	28.1	Salasad creek	Salasad Creek	A	A
Salasad 2	54.9	Salasad creek	farm ditch	>A	A

A - adequate   > A - more than adequate   L - limited   S - short

There was a spatial variation in the availability of irrigation water supply within the service area of Balanac RIS (Figures 9.8a and 9.8b). The upstream farms had greater advantage in securing their water supply than the downstream farms. Most upstream TSA and those along the main canal can divert their irrigation water practically anytime they want since the irrigation is continuous. This "quality" irrigation service to them was at the expense of the downstream farmers who had to wait for enough water to reach their canal before starting their farming activities for the season.The high dependence of water availability downstream to irrigation practice of upstream farmers was one of the perennial issues affecting farms at tail-end of the canal network. The cropping season in downstream farms usually starts between August and September for the wet season and between December to Janunary for the dry season. These starts of cropping are about three months behind of that of upstream farms. Such late starts result to the latter part of the crop growing period coinciding with the onset of unfavourable weather conditions - either well within the driest months or the typhoon season.

The two major creeks (Biñan and Salasad) traversing the service areas were tapped for irrigation purposed by installing check gates along their stretches. They are fed by drainage water from upstream farms, hence, are also dependent on diverted water by the dams, especially during the dry season.

In times of El Niño or a prolonged dry season, downstream farms are water-scarce or left idle due to lack of irrigation water. About 59 farmers used water pumps to augment their irrigation supply in about 52 ha farms downstream, 33 and 19 ha of which were supplemented by groundwater and lake water, respectively (Figure 9.9). Some upstream farmers can cultivate three crops in a year, while tail-end farms can only have a maximum of two.

Downstream farmers stated that such unequal access to irrigation water was attributed to poor flow control and laissez faire cropping and irrigation offtaking. While the irrigation supply is continuous a mix of non-functional and makeshift flow control structures, such as vandalized proportional and duckbill weirs and slab-type cross regulators, can be found along the main conveyance canals. A weir-type and orifice-type offtakes coexisted at the junction of the main canal and laterals Buboy and A. The former and the latter are located immediately upstream and downstream of long-crested weirs, respectively.

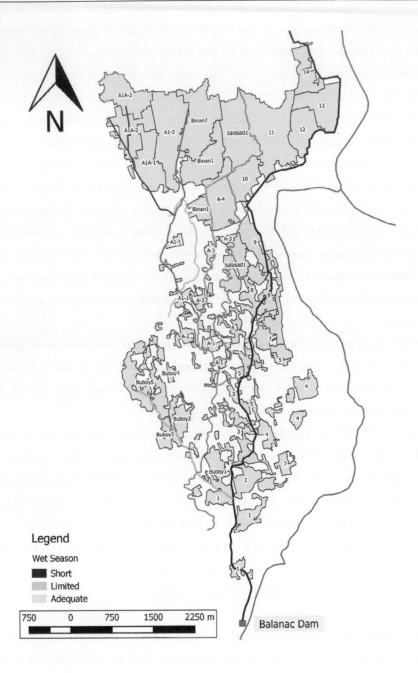

Figure 9.8a. Variations in water supply conditions in the different turnsout service areas of Balanac RIS during the wet season

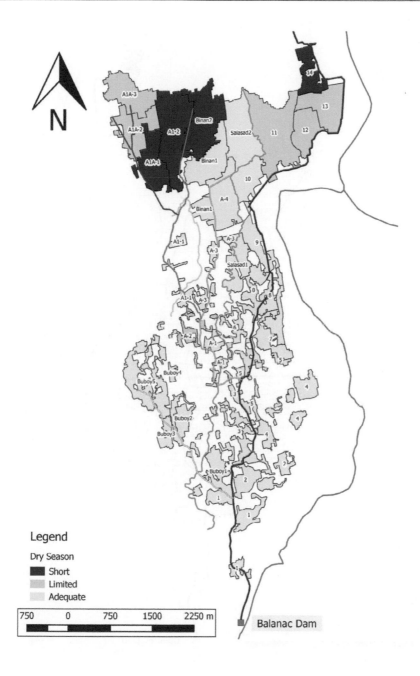

Figure 9.8b. Variations in water supply conditions in the different turnsout service areas of
Balanac RIS during the dry season

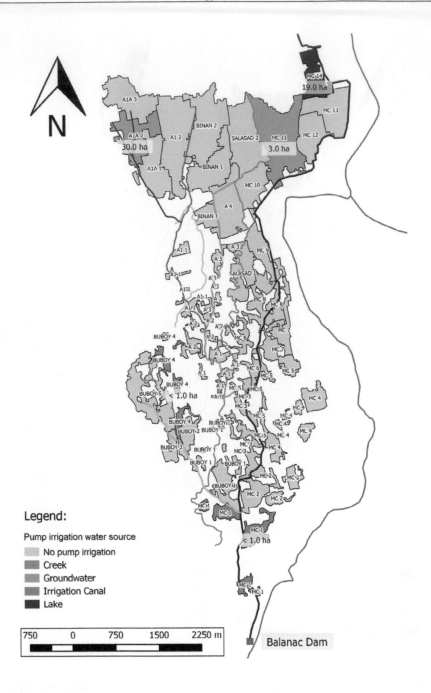

Figure 9.9. Turnout service areas of Balanac RIS with pump irrigation schemes

Many direct ungated offtakes further upstream are situated below the crest elevation of these proportional or duckbill weirs, giving them undue advantage in diverting the water. This could be the reasons why most fixed proportional weirs were eventually vandalized. With either vandalized proportional weirs or remnants of original vertical cross regulators at main distribution points, the ditch tenders and farmers either placed, or removed improvised check structures to modify water diversion. Consequently, the water delivery in the last two bifurcating canal reaches (Lat A1 and Lat A1A) needed close tending during the dry season to prevent illegal checking of water at upstream bifurcation points and turnouts. Unlike those along the upstream lateral (Buboy), water delivery to turnouts along these last two bifurcations was affected by flow situations in at least two layers of canal hierarchy. This means at least two layers of system dependency and longer flow path, hence more management resources.

9.3.2   Sta. Maria RIS

The service area of Sta. Maria RIS is partitioned into turnout service areas (TSA) based on hydraulic boundaries (Figure 9.10). Old system layout documents indicated that there were 65 TSA in the original system that covered a larger area. The present scheme has 30 TSA: 11 divert water from the main canal; 14 from secondary canals; 3 from both canals; and 2 are independent from the dam water supply, tapping drainage water, spring water, small streams, or a combination of these (Table 9.3). Each TSA is also represented by a farmer-leader.

Similarly, the Sta. Maria RIS has a relatively loose hierarchy of canals. Its main canal has more than 120 direct offtakes, about 80% of which are ungaged. Its longest lateral (A) has more than 70 direct offtakes. Similarly, about 80% of its direct offtakes are either ungaged or with removable metal flash board.

Unlike in Balanac RIS, the operation of dam gates and water distribution along the main canals were solely carried out by Santamasi, the water users association of Sta. Maria RIS. It mainly involved two of its officials: one for each dam and corresponding main canal. They operate the gates and effect water distribution plans themselves. The TSA leaders may only operate the gates of their turnouts following the instruction of the former. The gates of the main intakes are open most of the time and are closed only during occurrence of floods.

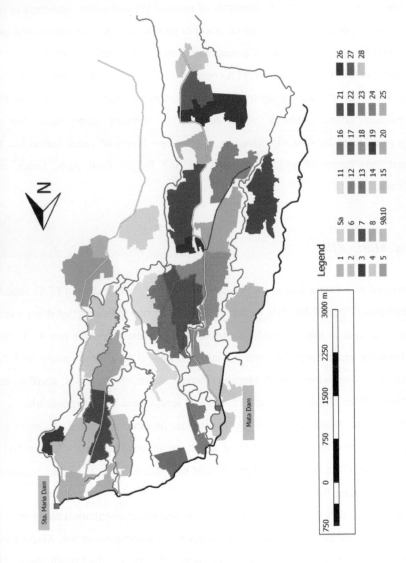

Legend

1	5a	11	16	21	26
2	6	12	17	22	27
3	7	13	18	23	28
4	8	14	19	24	
5	9&10	15	20	25	

Sta. Maria Dam

Mata Dam

750 0 750 1500 2250 3000 m

N

Figure 9.10. Turnout service areas of Sta. Maria RIS

Table 9.3. Profiles of Sta. Maria RIS tertiary service areas

TSA No	Service Area[1] (ha)	Water sources	Immediate supply canal		Percent of area supplied by the source	Water supply situation	
			Main	Lateral		WS	DS
1	27.0	Bagumbayan Dam	SMMC	--	100	A	A
2 & 3	55.6	Bagumbayan Dam		A	100	A	L
4	50.1	Bagumbayan Dam	--	A	100	A	L
5	41.2	Bagumbayan Dam	--	A	100	A	S
5A	33.6	Bagumbayan Dam	--	A	84	A	S
		Cabooan CG (Drain)	--		16	A	S
6	46.5	Bagumbayan Dam	--	A	50	A	S
		Cabooan CG (Drain)	--		50	A	S
6A	19.6	Bagumbayan Dam	--	A	50	A	S
		Cabooan CG (Drain)	--		50	A	S
7	44.3	Bagumbayan Dam	--	B	100	A	S
8	28.5	Pague CG (Drain)	--	B	100	A	S
9 & 10	32.5	Bagumbayan Dam		C	10	>A	S
		Masinao & Samala CG (Drain)			90	>A	S
11	41.5	Indocina CG		--			S
		- Salang Buaya (River)	--	--	36	A	S
		- Bucol (Spring)			36		
		- Samala, TSA 9&10 (Drain)			28		

Modernisation strategy for the national irrigation systems in the Philippines

Table 9.3 *Continued*

TSA No	Service Area¹ (ha)	Water sources	Immediate supply canal		Percent of area supplied by the source	Water supply situation	
			Main	Lateral		WS	DS
12	9.6	Bagumbayan Dam	SMMC	D	100	A	S
13	18.2	Bagumbayan Dam	SMMC	D	100	A	S
14	25.6	Bagumbayan Dam	SMMC	--	100	A	S
15	26.8	Bagumbayan Dam	SMMC	--	100	S	S
16	45.5	Mata Dam	--	E	100	A	S
17	54.5	Mata Dam	--	E	100	A	S
18	41.9	Mata Dam	MMC	--	100	A	L
19	20.8	Mata Dam	MMC	--	30	A	L
		Tagalag CG (River)	--	--	70		
20	43.3	Mata Dam	MMC	F	100	L	L
21	47.7	Mata Dam	MMC	--		A	L
22	48.3	Cambuja CG (Spring)	--	--	100	A to >A	L
23	25.6	Mata Dam	--	F	100	A to >A	A
24	31.1	Mata Dam	MMC	--	100	A	A
25	30.6	Mata Dam	MMC	--	100	A	L
26	43.95	Mata Dam	MMC	--	100	A	L
27	24.95	Mata Dam	MMC	--	100	L	S
28	12.2	Mata Dam	MMC	--	100	L	S

SMMC - Sta. Maria main canal   MMC - Mata main canal   CG - check gate   A - adequate   L - limited   S - short

Water distribution practice is rotation by lateral (secondary canals) and by TSA. The water delivery schedule among the TSA is agreed upon by TSA representatives and WUA officials prior to the start of the cropping seasons. Rotation starts at the most upstream laterals. Whenever an El Niño episode is expected to occur, irrigation rotation is started at the most downstream TSA clusters along the main canal. Strict implementation of the agreed-upon rotational irrigation schedule is followed during such a dry period. The tail ends of main canals, or those reaches immediately downstream of the last laterals are treated as laterals as far as irrigation rotation is concerned. For lateral A, the longest secondary canal, irrigation rotation is by group of TSA.

The permitted diversion schedule by direct offtakes or turnouts along the main canals is concurrent with the water delivery schedule of their respective TSA groups. The same rule holds for direct turnouts along lateral A. For the other laterals, their respective turnouts can simultaneously divert water. There are at most two levels of system dependence for most turnouts. It has been observed and concurred with canal operators and farmers that unauthorized diversions by ungated direct offtakes upstream of scheduled water delivery point were quite common. Such unpermitted diversions caused frictions among farmers and required extended man-hours of close supervision by the canal operators and, hence, management resources. They were undermining the implementation of agreed upon rotational irrigation.

The available irrigation supply differed within the service area of Sta. Maria RIS (Figures 9.11a and 9.11b). Seasonal variation in water supply was more notable than the spatial variation. Most TSA (with the exception of the most downstream areas) had adequate irrigation supply during the wet season. During the dry season, most farms (except the most upstream) had limited-to-short water supply.

Unlike Balanac RIS, five of the nine check gates for water reuse in Sta. Maria RIS are independent from canal network and dam water. They are fed by small creeks and natural drains originating from other sub-basins, dam tailwater or outflow, patches of springs along the foot of the mountain or a combination of these. Such water reuse check gates included Masinao, Samala, Indocina, Tagalag and Cambuja. The six TSA supplied by water from this check gates have the least dependence on system operation service (Figure 9.12).

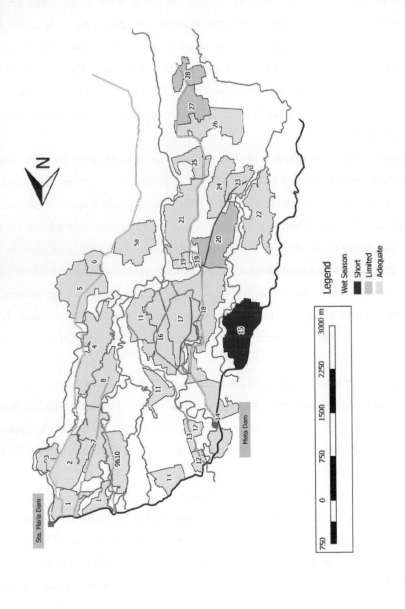

Figure 9.11a. Variations in water supply conditions in the different turnout service areas of Sta. Maria RIS during the wet season

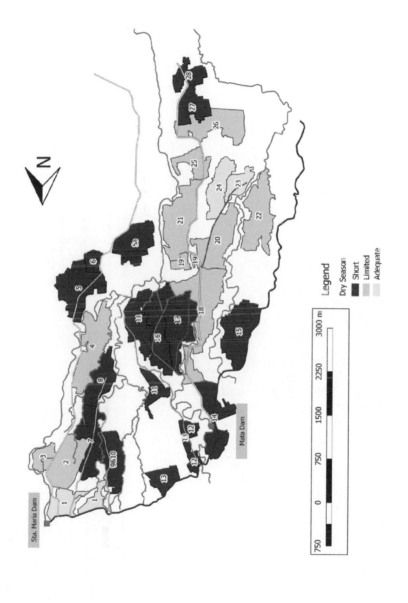

Figure 9.11b. Variations in water supply conditions in the different turnout service areas of Sta. Maria RIS during the dry season

Modernisation strategy for the national irrigation systems in the Philippines

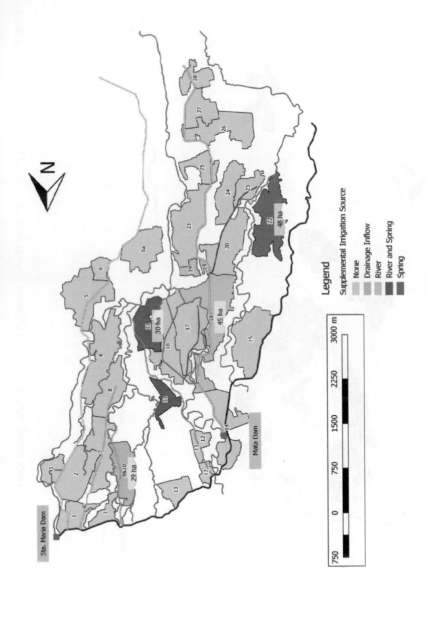

Options and vision for irrigation system modernisation plan

Figure 9.12. Turnout service areas of Sta. Maria RIS tapping supplemental water sources

9.4   Mapping the services and demand

On the average, the quality of irrigation service in the two case study systems was the same (Figures 9.13a and 9.13b). More TSA in Balanac RIS had a higher flexibility and reliability of irrigation service than those in Sta. Maria RIS. This can be attributed to the more abundant water supply and bigger canal flows in Balanac RIS. On the other hand, a more equitable water delivery service in Sta. Maria RIS could be mainly due to its strict implementation of irrigation rotation and similar water distribution schemes. However, it did not result to a better irrigation service because of the inherent inadequacy of irrigation supply, especially during the dry season. The readily apparent ways to improve the quality of irrigation service in Balanac RIS and Sta. Maria RIS lie in the improvements on water distribution and increasing the irrigation water supply, respectively. The demand for canal operation would be higher in Balanac RIS mainly due to the lack of functioning flow control structures and the higher proportions of ungated turnouts along its main conveyance canals (Figures 9.14a and 9.14b).

The most downstream farms of Balanac RIS and Sta. Maria RIS are subjected to seasonal flooding (Figures 9.15a and 9.15b). Flood inundation is more frequent in Balanac RIS where a significant stretch of its service areas are adjacent to the Laguna Lake. About 10 of its TSA experience destructive flooding almost on a regular basis. Majority of these are the same farms that had to wait for their upstream counterpart to satisfy their irrigation requirement and free up irrigation water they no longer need. Shallow tubewells and open source water pumps are used in at least three of these TSA to supplement canal water supply.

Except in the most upstream service area (Magdalena) the service areas of Balanac RIS and Sta. Maria RIS are well within the area classified by the National Water Resources Council (1982) as shallow well areas or areas with static groundwater levels within six metres below ground surface (Annex L). As evidenced by use of tubewells by farmers in downstream areas of Balanac RIS, there are potentials for shallow tubewell irrigation in these areas, which would help address the high dependency on canal water use upstream. Similarly, the number of free-flowing wells and ponded areas fed by spring water located in the vicinity of Sta. Maria main canal along the foot of the mountain range would indicate good potentials of these groundwater resources as supplemental source of irrigation water. There could be

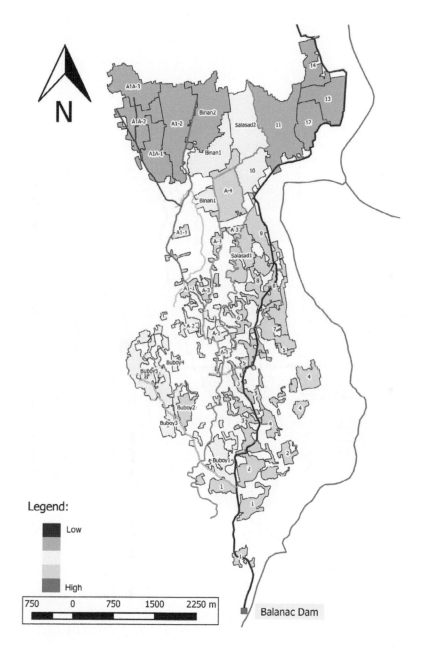

Figure 9.13a. Relatively quality of water delivery service in Balanac RIS

Modernisation strategy for the national irrigation systems in the Philippines

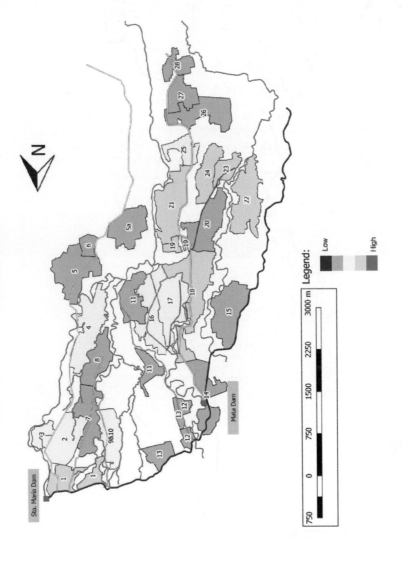

Legend:

Low

High

| 750 | 0 | 750 | 1500 | 2250 | 3000 m |

Figure 9.13b. Relatively quality of water delivery service in Sta. Maria RIS

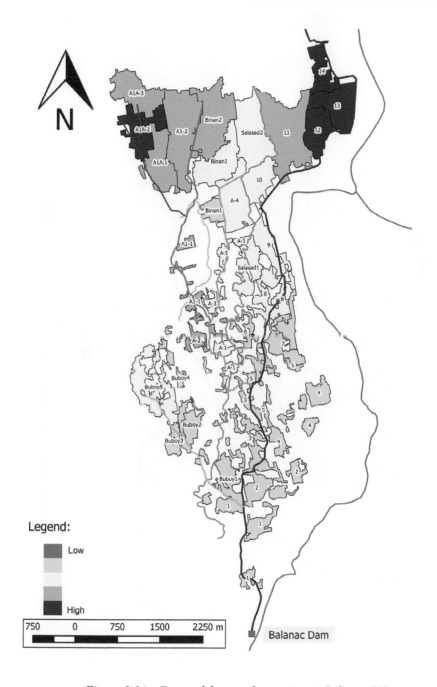

Figure 9.14a. Demand for canal operation in Balanac RIS

Modernisation strategy for the national irrigation systems in the Philippines

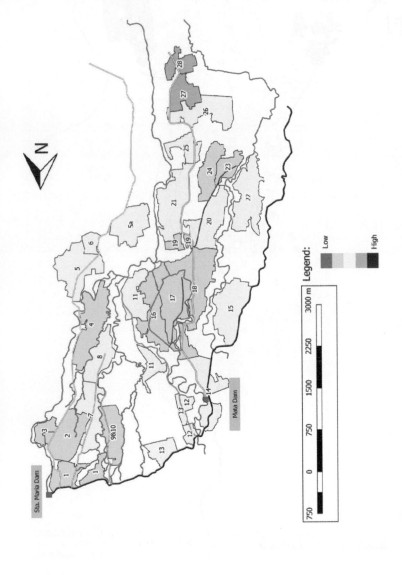

Figure 9.14b. Demand for canal operation in Sta. Maria RIS

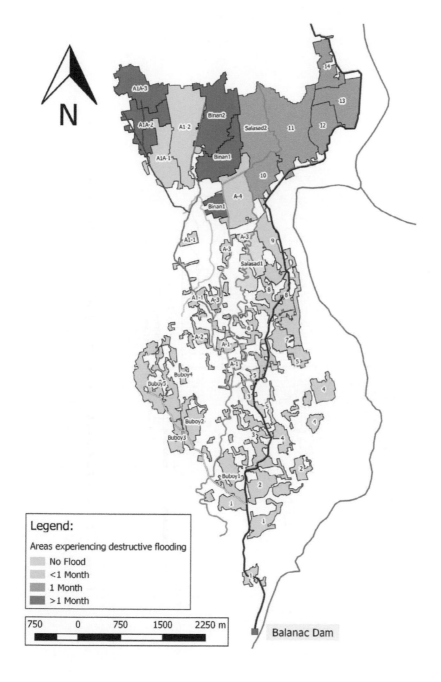

Figure 9.15a. Balanac RIS turnout service areas affected by seasonal destructive floods

Modernisation strategy for the national irrigation systems in the Philippines

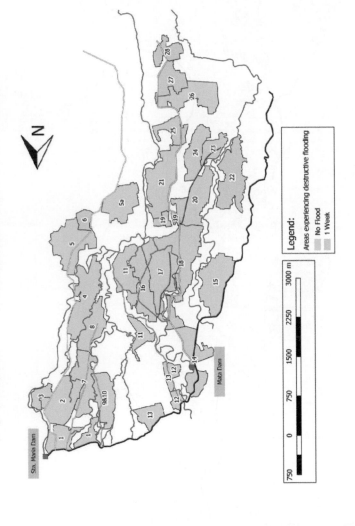

Figure 9.15b. Sta. Maria RIS turnout service areas affected by seasonal destructive floods

sites in the downstream areas of Balanac RIS (Pagsanjan and Sta. Cruz) and Sta. Maria RIS where hand tubewells would be the more appropriate scheme as indicated by the low specific well capacities obtained by the NWRC (1982), ranging from 0.3 - 2.0 lps in downstream areas of Balanac RIS (Pagsanjan and Sta. Cruz) and from 0 - 3.1 lps in sample sites in Sta. Maria RIS, would suggest that hand tubewells are appropriate for use in these areas.

Both Balanac RIS and Sta. Maria RIS were designed and constructed to provide water for rice monocropping. Though still mainly for rice irrigation, there were other water uses that have emerged within the service areas. These include fish ponds, aquarium fish nursery, piggery, poultry, orchard and vegetable farms (Figures 9.16a and 9.16b). They represent additional water demand and different delivery service requirements. At present, the use of irrigation water for purposes other than rice irrigation is discouraged.

9.5 Field investigation of potentials for conjunctive use

Well drillings were carried out in two downstream sites in Sta. Maria RIS: one was situated at 14.50835° N, 121.42311° E adjacent to Sta. Maria main canal in the vicinity of TSA 13; and the other at 14.44339° N, 121.41962° E in TSA 28 downstream of Mata main canal (Annex M). There were at least one existing tubewell in the neighborhood of each of the selected drilling sites. However, the drilling activity was not successful in both sites due to unfavorable formation underneath.

The drilling site near TSA 13 was underlain by loose rocky formation to at least about 15-meter depth as indicated by the well cutting materials obtained from the borehole. The drill head (drilling stem and drill bit) detached from the rest of the drilling stem and got stuck in the formation.

No good aquifer was found in the drilling site near TSA 28 within the almost 55-meter drilling depth. The site was underlain by a clayey soil formation as suggested by the well cutting materials from the well borehole. The drilling activities for the two sites were stopped since the chance of hitting an artesian well that would allow pumping water out by suction lifting got slim.

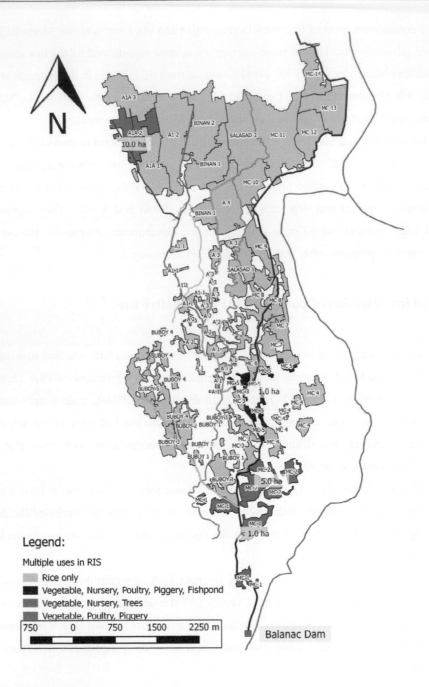

Figure 9.16a. Homogenous rice irrigation demand and emerging water uses in Balanac RIS

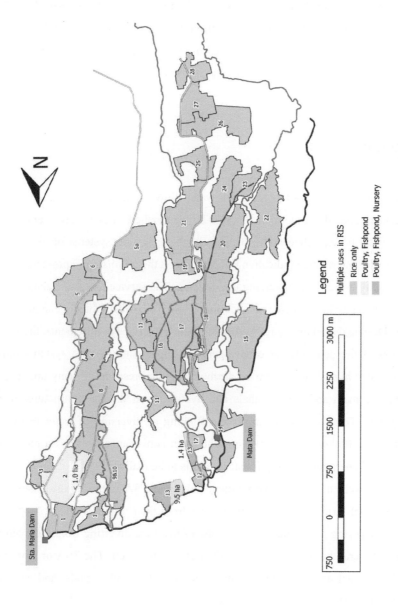

Figure 9.16b. Homogenous rice irrigation demand and emerging water uses in Sta. Maria RIS

This drilling experience would suggest a presence of ground formation classified as difficult areas within the service area of Sta. Maria. A geo-resistivity survey of the area would be useful in this case. Pumped irrigation would still be feasible in difficult areas, but tubewell drilling and development would be relatively difficult as ground formation in these areas are characterised by sheared rock formations. Based on NWRC (1982) estimates the probability of occurrence for deep well sites in difficult areas are 50% while that of shallow well sites and non-productive well sites are 25% each.

9.6 Conclusions

The recommended characterization steps of water flow path, management units, O&M, irrigation service and demand in MASSCOTE compelled a structured and oriented examination of the present state and prospects of the irrigation systems on these aspects. Thus, they provided interrelated indicative information on the promising improvement options and course of action towards modernizing the systems. The service areas of Balanac RIS and Sta. Maria RIS are cultivated for rice monoculture and have basically homogeneous irrigation requirements. Despite the relatively relaxed water delivery service requirements, the results of characterization of the demands for canal operation supervision in both systems indicated high demands. Such high demands can be attributed to the presence of many ungated direct offtakes along the main and lateral canals and dysfunctional diversion structures at major distribution points. There should be no direct offtaking or diversion of water to individual farms and ungated offtakes or turnouts along the main conveyance canals if a more orderly and equitable canal operation and water distribution is to be realized.

While there were delineated management units and TSA organizations, many have a number of individual farm turnouts diverting water from their respective TSA supply canals which are also supplying downstream TSA. Some of them are diverting from the main canal. Such loose canal hierarchy resulted to unwieldy water distribution. The TSA or management unit should have equal access to their common immediate supply canals, and as much as possible, have only one diversion point for each. This would help consolidate each TSA for a more manageable system operation and strengthening WUA leadership in carrying out water distribution duties. Limiting the number of diversion points to the optimum would require revisiting farm canal layout in each TSA and ensuring each farm has access to a farm ditch.

There were spatial variations of water adequacy within the service areas of Balanac RIS and Sta. Maria RIS. Water deprivation and submergence of tail-ends TSA were experienced in both systems, especially during El Niño and La Niña years. Meanwhile, composite maps of the water flow paths showed potential sites for additional water reuse facilities. Also, the presence of shallow aquifers in downstream areas of Balanac RIS and patches of springs in Sta. Maria RIS indicated potential sources of supplemental water for irrigation. As demonstrated in Sta. Maria RIS, implementing a system-wide cropping calendar, hence, rotational irrigation schedules for each TSA helped in coping with their limited water supply, but only to a limited extent. It could not circumvent the water supply shortage caused by strong El Niño events. Also, their practice of shifting the planting season ahead of the onset of rainy months to avoid the destructive period of the monsoon season was constrained by the low available water supply from the rivers. Planting at these earlier dates is more favourable only if irrigation water is available. The groundwater resources in the service areas present an alternative source of additional water. Shallow tubewell irrigation development in tail end areas of Balanac RIS and the feasibility of spring development for irrigation supply augmentation for Sta. Maria RIS would have to be explored.

The results of the expenditure analysis showed that the cost of frontline O&M activities in both systems were below the PhP 2,000 - 3,000 ha^{-1} standard planning budget and the PhP 1,900 ha^{-1} actual average expenses for O&M nationwide. The WUA budgets, which constituted the amounts spent for field system operations and were sourced from ISF collection, were much lower than the amount of ISF collected in respective systems. For the 2010-2015 period, they averaged at about 53% and 38% of the ISF collected in Balanac RIS and Sta. Maria RIS, respectively. It should be noted that Sta. Maria RIS received a lower percentage of its ISF collection despite its slightly higher ISF collection (PhP 2.2 million) than that of Balanac RIS (PhP 2.0 million) and the same stage of IMT as Balanac RIS.

The two systems spent about 70% of their respective WUA budgets on wages and salaries of its field personnel. Only about 12% and 6% were spent for canal repair and maintenance in Sta. Maria RIS and Balanac RIS, respectively. While there were some cost items where savings are possible such as miscellaneous in Balanac RIS and travel in Sta. Maria RIS, an increase in the current WUA budget would have to be considered by the NIA. Meanwhile under the current arrangement, the WUA budget increases with increased ISF payment. Thus, mechanisms or programs to encourage or enforce ISF payment and collection would have to be initiated by the NIA and WUA.

While the service areas of Balanac RIS and Sta. Maria RIS remained for rice mono cropping, a few other agriculture-related uses have emerged within the service areas such as aquaculture, livestock and vegetable farms and orchards. As traditional rural community development multiple uses of water can be expected. Likely-possible future scenarios on multiple uses of water may have to be integrated in irrigation system modernization plans.

10 OPTIONS AND VISION FOR IRRIGATION SYSTEM MODERNISATION PLAN

10.1 Introduction

Possible ways to improve the system operation, maintenance and management, water delivery service and overall system performance would become evident as in-depth analysis of the different aspects of an irrigation system move through a series of impact assessment, system diagnosis, design revalidation and system characterization studies. The feasibility and appropriateness of such potential management and/or technical solutions or modernisation options would have to be examined at both the level of their respective management units or turnout service areas and at the main system level. Modernisation options identified at the local TSA level must be doable and agreeable with the management and operations at the main system level, and vice versa.

How the system officials and farmer-water users envision their desired future scenarios for the irrigation system and services in the context of foreseeable reality would set the targets and general direction for formulating an irrigation system modernisation strategy. Also, such vision would have to serve as the basis in selecting or developing a set of relevant modernisation solutions among the possible options. It is with a consolidated vision for an irrigation system that the formulation of a modernisation plan needs to start.

This chapter enumerates the promising management and technical modernization activities that have become evident from the findings of the research study. It discusses a strategy for modernising the Balanac RIS and Sta. Maria RIS and other national irrigation systems that are in a similar situation.

10.2 General methodology

Potential improvements of the physical strutures, operation, maintenance, management and institutional aspects for the case study systems and national irrigation systems in general as revealed by the findings of the course of studies carried out were collated for consideration for inclusion in a modernisation plan. The mental images of the desired future state of their respective irrigation systems as perceived by the NIA system officials and farmer-water users were solicited through consultative meetings with them. Open-ended questions on how they would like the future state of their irrigation systems, quality of irrigation service and irrigated agriculture in their municipalities to be like in five to 10 years from today were fielded to facilitate the start of the vision formation activities. These also included the constraints and potentials to improved irrigation system performances and what they can contribute to a modernisation program for their systems.

Outlines of a modernization plan for the case study systems were drafted based on findings of the assessment, diagnostic and characterization studies as well as based on the articulated vision by the system officials and farmer-water users for their respective systems. This can serve as working documents for preliminary consultative meetings or planning workshops for an irrigation system modernisation program for the case study systems in the near future.

A strategy for developing a modernization plan for similar national irrigation systems in the country was formulated based on the procedures and tools employed in the case study systems. Central to the modernisation strategy is the foremost importance of a coherent link of the system design, operation, maintenance and water supply to improve irrigation water delivery.

10.3 Modernisation options for Balanac RIS

10.3.1   System management

The following modernisation options for the management of Balanac RIS have been formulated:

- *adoption of coordinated, system-wide cropping and irrigation delivery schedules.* Laissez faire planting schedules and diversion of irrigation by upstream farmers disadvantaged downstream farmers in terms of ready access to irrigation water and a lost opportunity to strategically schedule their farming activites to avoid typhoon-and drought-prone months. While upstream farmers can start their land preparation at their preferred dates, the downstream have to wait until upstream farmers do not need much water before they can start cultivating their farms. Well-defined and agreed-upon cropping and water delivery schedules among clusters of turnout service areas or management units would promote equity in access to water, orderliness in water distribution and increase in irrigation coverage in downstream areas;

*   *adjustment of the cropping calender to earlier planting dates.* The downstream portion of the service areas of Balanac RIS bordering the Laguna Lake are subjected to perrenial floods or inundation. The actual service areas irrigated were smaller during the wet season mainly because a significant portion of downstream service areas was submerged. Typhoon and destructive monsoon damages were the main reasons for lower yield and more ISF payment exemptions. Shifting of planting season to an earlier period would help to avoid the destructive part of the monsoon season;

*   *downstream-to-upstream irrigation schedule.* Cultivation and irrigation water delivery for the wet season or a crop year can be started at downstream areas then gradually going upstream so that farms or crop submergence can be avoided;

*   *conjunctive use of shallow tubewells (STW) in downstream areas.* The use of small, individual pumps for irrigation were in recent years not promoted by the system officials to prevent issues on ISF payment exemptions for those farmers who pumped their own irrigation water and reduction in collectible ISF. With a more frequent prolonged drought and lower dry season river flows, alternative modes of irrigation like tubewells would have to be supported as part of the coping mechanisms of irrigated agriculture to El Niño-induced drought. The irrigation agency can initiate aquifer characterization in the area to determine sustainable pumping rates and population density of STW;

*   *canal operation supervision.* Water distribution and delivery were carried out by the two WUA ditch tenders who were deemed without authority in water distribution functions and hence, whose daily canal tending duties are dismissed or overruled by most water users. Canal operation would need to be under the supervision of persons elected or consented by the water users to be the authority so that the implementation of agreed-upon water distribution and delivery schedules may be respected by them;

- *minimize flow dependence of sub-lateral canals.* Availability and accessibility of irrigation water for turnout service areas served by Laterals A1 and A1A are dependent on flow situations of at least two levels of canal hierarchy. Operational adjustments for the conveyance losses and travel time of allocated discharge for these canals would have to be incorporated in the water distribution plan. Lateral A and Lateral A1 would have to be considered as part of the main conveyance canal system to level off system dependence of the major distribution canals;

- *water distribution plan and gate operation manual.* While fixed proportional long crested weirs were installed at the diversion point of the first two laterals and in some lower level canals, makeshift check structures were sometimes also used to divert water into these laterals. Rotation by lateral was also practiced during times of low river flows. Irrigation rotation and makeshift check structures were employed on an ad hoc basis and often unilaterally by system operation personnel. An official water distribution plan and canal operation instructions would have to be developed in consultation with the turnout service area leaders and dissiminated prior to the start of a crop year and each cropping season;

- *dam desilting.* Almost the whole length of the Balanac dam as filled with mixed of rocks, boulders, silt and other river sediments up to the crest level. As a result, the headwork was reduced to somewhat like an intake barrel type structure. The sluice gates designed to flush sediments and prevent them from flowing into the main intake are used to raise the water level near the main intake and effect diversion by gravity. The dam would have to be desilted to restore its diversion capacity and limited flow modulation capacity;

- *flow measurement.* Measurements of available water supply and diverted water have to be carried out to collect flow data that can be analyzed to support the seasonal programming of the areas to be irrigated and the corresponding budget and resources allocation. A more realistic system operation, maintenance and management plan and projection of additional service areas and areas that can be restored or modernised would be possible with analysis of actual flow data.

10.3.2   Physical system

The following modernisation options for the physical system of Balanac RIS have been formulated:

- *consolidation of turnouts along the major conveyance canals.* The total number of turnouts along the Balanac main canal was more than double their original count. Most of the additions were unofficial turnouts. Each turnout along the major conveyance canals represents a unit workload for canal operations and management. The existence of turnouts along the main canal needs to be rationalized based on hydraulic considerations. Restoration of the original and construction of additional main farm ditches would have to be carried out to reduce the many turnouts to an optimal number;

- *repair of canal embankments.* The canal embankments of the main canal at several locations need to be compacted to stop considerable water leakage through them so that conveyance losses can be minimized and irrigation water can be delivered within the target rates and duration;

- *consistency in system operational objectives.* The continued use of the original orifice turnouts and check structures at the bifurcation points for upstream laterals and at some downstream diversion points ran counter to the operation objective of a splitted flow method of water allocation to laterals and splitted flow method of water distribution through the main system apparently adopted in the Balanac RIS as manifested by long-crested, fixed proportional weirs. Also, the presence of ungated direct offtakes below the crest elevation of the proportional weirs immediately downstream undermines the purpose of the proportional flow control method. The flow control structures need to be consistent with the goal of the chosen canal operation and water distribution;

- *repair or replacement of disfunctional flow control structures.* A fixed, long-crested proportional weir that replaced the original vertical gates was vandalized with a section of its crest destroyed. Contemporary duck bill weirs along minor conveyance canals were also vandalized. The original gated turnouts and check structures at major bifucation points still exist in their place without their gates and other movable parts. These lack of functional flow control structures resulted in unwieldy water distribution and chaotic water delivery. Socially acceptable and a solid hydraulical combination of flow control structures would have to be adopted for a more orderly water distribution and equitable water delivery service;

- *repair of the dam intake gate and sluice gates.* The mechanical lifting mechanism and the metal guides of the dam sluice gates would have to be replaced by new ones to facilitate ease in dam operation. The bended steel frame of the intake gate needs to be repaired or replaced so that it can be fully closed during times of big floods to prevent entry of flood water and submergence of the service area;

- *construction of a reservoir-type dam.* The run-of-the-river diversion dam has inherently limited capacity to support productive irrigation for rice monoculture in times of prolonged drought and low river flows. Construction of a reservoir-type dam to support irrigated agriculture is increasingly becoming a necessity to cope with climate change;

- *second Balanac dam at Lower Balanac River.* Considerable volume of water supply just flows over the diversion dam especially in cases where the dam is heavily silted. Such water can also be diverted at some downstream river locations within the vicinity of the midstream-to-downstream service area. This will minimize system dependence of dowstream areas and decouple downstream canal operation from that of the upstream.

10.4 Modernisation options for Sta. Maria RIS

10.4.1 System management

The following modernisation options for the management of Sta. Maria RIS have been formulated:

- *flow measurements.* The approach sections of the flumes, which are used for gauging whether the diverted flows would allow relatively relaxed water distribution schedules or require implementation of a tight irrigation rotation, need to be desilted and restored to their ideal conditions for a more accurate flow estimation. Staff gauges for these main intake flumes need to be reinstalled and the respective discharge rating curves revalidated. Similarly, flow measurement structures at major distribution points have to be revived to enable monitoring of incoming supply and diverted water for more transparent and objective rules for water distribution and verifiable irrigation service delivery;

- *desilting of dams.* The Bagumbayan and Mata dams could only raise the surface elevation of a limited volume of river flows as they were heavily silted that bushes were growing on the island of river soil created at the upstream approach of the dams. The upstream approaches of the dams need to be dredged to restore the diversion and the inherent storage capacity of the dams.

10.4.2   Physical system

The following modernisation options for the physical system of Sta. Maria RIS have been formulated:

- *repair of dam intake and sluice gates.* The sluice gate of Bagumbayan Dam remained fully closed and inoperational during at least the past 10 years. It would have to be replaced by a new one so that entry of sediments and rocks into the canal network can be avoided. Meanwhile the flushboard-covered sluice gate of Mata Dam is inherently insufficient to effectively flush out sediments and risky to adjust during times of flood when it is best left open. The intake gates of both dams need repair and/or replacement;
- *repair or replacement of disfunctional flow control structures.* Several major turnouts and check structures along the main canals are either damaged or missing their gates. The lack of functional gates results in insufficient control of canal flows and difficult management of water distribution. Morever, it makes supervision of water distribution cumbersome as unauthorized gate opening or adjustment can be easily done with makeshift flow control structures, hence, the need for frequent rounds along the canals by the operators. Disfunctional and missing gates need repair or replacement by new ones to improve flow control and to achieve the water delivery targets;
- *consolidation of turnouts along the main canals and lateral A.* The number of official turnouts along these canals where irrigation rotation by service areas was practiced accounted for about 50% of the existing turnouts along these canals. The total number of turnouts needs to be reduced to an optimal number based on hydraulic considerations to simplify and more efficiently manage canal operation and water distribution.

Additional main farm ditches that can serve several neighbouring turnouts can be constructed so that integration of such loose turnouts into a more well-defined canal hierarchy would be possible;

- *closure of ungated direct offtakes along major conveyance canals.* The presence of direct offtakes along the main canals and lateral A would undermine an orderly implementation of irrigation rotation. Ungated direct offtakes would need to be closed permanently to prevent any unauthorized diversion that causes chaos in achieving the water delivery targets, and consequently, conflicts among water users and canal operators;

- *reservoir-type dam.* The segmental relay cropping and irrigation rotation practices, which have been adopted to cope with the limited water supply, may have sustained the perennial presence of pest and diseases that resulted to lower yields in several turnout service areas. The irrigation system would need reservoir-type dams, which would enable storing water during high river flows and making it available for use especially during the dry seasons;

- *spring development.* The several springs found along the foot of the mountain bordering the service represents potential sources of additional irrigation water for the irrigation systems. The feasibility of developing these groundwater supply sources to augment the irrigation supply from the dams would need to be explored;

- *Construction of structures for drainage water reuse.* Drainage water flowing near TSA 18 and 19 presents an opportunity to augment water supply for these areas. Additional check gates would need to be constructed to capture this water for irrigation purposes. Drains would need to be constructed to facilitate flows of excess water into natural waterways crossing the service area and, eventually, to strategic locations for water reuse structures (check gates).

10.5 Visions for system modernisation

In general, system officials, field personnel and WUA leaders of Balanac RIS and Sta. Maria RIS have hoped that an entire firm up service area (FUSA) will be adequately irrigated in the near future. For the NIA system officials and field personnel full irrigation of FUSA would indicate an exemplary system and management, increase ISF collection and a concrete contribution to the government goal of increase in rice production. For the water users it would mean more harvest, increased income and a better life.

10.5.1   Balanac RIS

The current system and WUA officials were not certain if there were a stipulated vision, mission and objectives for Balanac RIS as they do not have a copy of such document. A vision could be drafted based on the results of the WUA consultation-workshop on their dreamt and desired state for Balanac RIS (Annex N). The general picture emerging was that of a strong, progressive and well-governed WUA with active involvement in water distribution, system operation and management and planning for system improvements to provide equitable, orderly and quality irrigation service. About two-thirds of the WUA officials and TSA leaders have envisioned the following features for their system:

*   *equitable water distribution.* Recognition of equal rights of members to water and irrigation service, agreements to allocation of water based on farm size, adequate water for all members, and approval on the implementation of irrigation rotation and system-wide cropping calendar, including cropping schedule adjustments to cope with adverse weather;

*   *strong and progressive WUA.* Desires for cooperation among members, a more transparent system governance, a quality irrigation service to achieve at least two croppings and a more active TSA leaders participation in planning system administration, physical structure improvements and water distribution;

*   no-leak, well-maintained and concrete-lined canals.

Fifty-five to 60% of them expressed desires to have the following: legal protection for the dam against ongoing quarrying operations, operational gates for the dam, undamaged canal flow control structures, canal spillway for excess water, additional check gates for drainage water reuse, flow measurement structure for the main intake, storage-type dam, reliable water delivery schedule, written rules on opening of new farms and penalty for noncompliance to water distribution rules.

While most participants have envisioned equitable allocation, quality irrigation service and irrigation rotation, lesser have envisaged water supply and canal structures aspects to achieve them. With the water supply and headwork aspects not among the top most desired system features, it can be said that the majority of farmers do not see them as a major or urgent problem. Such view could be attributed to the adequacy of available canal water in upstream areas and to the fact that seasonal water scarcity is experienced in downstream areas only. Meanwhile though the majority liked irrigation rotation schemes to be implemented, there was a divide among WUA members on the choice between "with adjustable gates" canal flow control structures and "without adjustable gates". Also, there was not only a split among them in the case of direct turnouts along the main canal, but also a low response (about 40% of the participants) on this matter. Such apparent disconnect of the desired general idea to the specific mechanism to realize them could be attributed to doubts or less clear understanding on which physical structures and distribution schemes would lead to the envisioned general scenario.

10.5.2   Sta. Maria RIS

A written vision, mission and objective for Sta. Maria RIS is available at the office of the WUA. The vision stipulates a strong association with united and active members in pursuit of better lives for themselves and in their community. The mission is to provide the irrigation service required to enrich farmers so that they may secure good education for their children and contribute to community's progress. The stated objectives include the following: to broaden their member's knowledge on modern farming technology, to sustain the WUA vision by implementing policies supportive of improved and orderly irrigation system, and to help members have income and livelihood through farming.

While the above articulated vision, missions and objectives are in general terms, a more specific description would serve as a better guide in the formulation of an irrigation system modernisation plan. The results of the consultation workshop carried out painted the main feature of the WUA-envisioned future Sta. Maria RIS: a water secured irrigation system managed and operated by a strong, self-sustaining and progressive WUA and capable of delivering the required irrigation service throughout its service area, non-rice agriculture and neighbouring rainfed rice areas. The most popular (preferred by at least around two-thirds of TSA leaders) of the more detailed descriptions of their desired future scenarios (Annex N) included the following:

- WUA has good governance, capability for planning and implementing system rehabilitation, maintenance and improvement works, has reached IMT stage 4 (complete turnover of management, operation and maintenance of irrigation systems to WUA) and a viable government partner in the implementation of agricultural development projects;

- increased irrigation supply through fully functioning and well-maintained headworks, construction of additional water reuse structures (check gates), installation of spring development structures and, eventually, construction of storage-type dams;

- flow measurements along the canals and flow gauging of the rivers;

- equitable and orderly water distribution as indicated by preference of water allocation based on farm size and equal rights of each member, legal and authorized turnouts only, construction of sufficient number of main farm ditches to minimize direct offtaking along main canal and laterals, and having sufficient, well-functioning flow control structures;

- climate-proactive cropping calendar and water distribution schemes (irrigation rotation, early planting dates, downstream-to-upstream-farms water delivery schedule) supported by flow control structures with adjustable gates;

- reliable irrigation service as suggested by the desires to establish water rights of the original service area, facilitate canal flows (desilted, well-maintained, without leaks, lined main canals) and have structures that are amenable for timely delivery (intermediate reservoir, main canals with storage capacity).

The above list of desired scenarios suggested that a sizeable majority of WUA representatives had clear knowledge of their most pressing concerns on their irrigation system and were more united on how they would like to address them. Unlike that of Balanac RIS, WUA repsentatives of Sta. Maria RIS had a clearer vision of their future modernised system. Such coherence and specific choices in the envisioned system could be attributed to their mutual experiences and struggles related to shortcomings in their present irrigation system and to their more active involvement in running the system.

10.6 Development of a modernisation plan for NIS

The NIS in the country differ in terms of type of headworks, size of service area, vintage, irrigation technology, management setup, level of development and physical environment, among others. While these various types and categories would seem to require as many varying solutions to improve the systems performances, a general strategy for developing an irrigation modernisation plan can be formulated. Central to this strategy is to critically analyze the link among system design, operation and water supply so that potentials and specific solutions for system modernisation can be identified.

The proposed strategy for formulating a NIS modernisation plan starts with an overview of the state of irrigation system performance based on a common and readily available indicator - actual irrigated areas. Provision of irrigation to its design service area is a common, basic goal of an irrigation system. How well this goal is achieved, that is, what percentage of the design service areas was irrigated would be indicative of the system performance.

The primary determinant of the extent of areas that can be irrigated is the available water at the source. Thus, the adequacy of available water supply to irrigate the design service area needs to be investigated also. In cases of ungauged water supply source and irrigation system, local knowledge on water availability for the system and an analysis of the historical data on actual area served vis-a-vis irrigation service areas would lend indicative information on possible constraints to achieving 100% irrigation coverage and potentials for expanding the irrigated areas.

The third step of the proposed strategy is the assessment of the nature and impact of intervention measures carried out to maintain the service area, expand the area actually served and to improve the water delivery service. Understanding the nature of rehabilitation and/or improvement works and their impacts in terms of either maintaining the irrigation service areas, expanding irrigated areas or improving irrigation service would reveal to what extent the different intervention measures have been carried out and which measures would have good potentials for achieving the irrigation system targets.

Irrigation systems can perform only up to their design capacity. It would be unreasonable to expect a level of system performance that goes beyond what the physical system is made capable of. They have to be operated in such a way they are designed to run.

Otherwise, there will be chaos and frustration in the system. With an overview of level of system performance, adequacy of available water and efforts to maintain or improve the irrigation service, a more comprehensive system diagnostic need to be carried out so that root causes of mediocre performance and specific solution can be identified. System diagnosis is especially important in the case of most NIS since they have undergone a mix of technical, management, policy and institutional changes whose interplay and effects needs to be systematically analyzed to be well understood. The strategy proposes a diagnostic assessment framework that links NIS performance to logical coherence among the scheme objectives, physical structures, management and water supply. It uses a combination of the logic design framework by Ankum (2001), diagnostic tools of MASSCOTE, discharge-head relations and hydraulic flexibility concept (described in Chapter 7). The headwork components which are lacking in both the logic design framework and diagnostic tools of MASSCOTE was included to put the water supply concerns in clear perspective.

Some of the design values and ensuing canal flow hydraulics may have been based on generalization or certain assumptions during the planning stage of irrigation system development. Designing for system improvement requires location-specific data and a more reliable estimate of design criteria. Thus, collection and revalidation of values of design parameters form part of the necessary undertakings to support a foreseable modernisation program. Revalidation of the percolation rates and assumptions on water availability are deemed most pressing in the case of the NIS, especially for the ungauged systems. Heterogeneity in soil types may be observed even in small-scale systems, hence, different percolation rates. Even in monocropped areas, this would mean different irrigation requirements that would need to be taken into account when planning for improvements in operation and water management schemes and designing for the corresponding flow distribution structures. Meanwhile, changes in riverflow regimes have been generally observed and were attributed by many to variability in weather patterns and in land use changes within the river basin. River flow gauging and canal flow monitoring needs to be started in several ungauged river basins and irrigation systems in order to have build a more reliable flow data set for planning a more rational modernisation programs. Thereafter, assessment of water availability by using appropriate modern flow simulation models would be more sensible.

The proposed strategy adopts the general concepts of the MASSCOTE methodology of mapping or characterization of water flow paths, O&M, irrigation service, unit service areas, and demand for operation. Whenever applicable and data are available the MASSCOTE procedures were followed. Otherwise equivalent or applicable, related analyses in the case of NIS were carried out.

Participatory approach to formulating a vision for an irrigation system would not only promote a sense of responsibility for and ownership of the system among farmers but also help ensure that the ensuing modernisation plan includes managerial and technological improvements that are coherent, technically feasible and socially acceptable. Irrigation designers and practitioners have this crucial role of extending techical knowledge so that farmers make informed choices and decisions concerning their future irrigation system. Relevant and farmer-appropriate trainings on irrigation system operation and water management, logical combination of flow control structures and water distribution schemes, coherent general system design for envisioned modernised irrigation systems, among others, may be initiated by institutional development personnel of the NIA to support irrigation system modernisation programs.

10.7 Conclusion

Though with basically the same general vision for an irrigation system such as attaining 100% area coverage of water delivery service for the two cropping seasons, water users could be found either electing contrasting choices on or being mum about the specific modernisation methods. This was observed in the WUA of Balanac RIS. An opposite situation was found in the WUA of Sta. Maria RIS, where there was clear consensus on the general and specifics terms on desired features of their future modernised system. Such cases could be explained by the more active involvement of the latter in running their system. Similarly, while objective, independent analysis of modernisation prospects for a system could come up with a comprehensive draft of a modernisation plan, it has to harmonize such draft with the specific modernisation features preferred by the water users. Participatory approach to system modernisation planning would be more fruitful if water users are equipped to make informed choices.

The options to modernise the case study systems related to both management and physical aspects. From financial point of view, modernisation solutions would cost much less than those concerning the physical improvements. However, they might have to weather more resistance, especially from the upstream farmers of Balanac RIS who would have to adjust their farming and irrigation practices. Winning the upstream farmers' agreement to these status quo changing modernisation options would need an extensive educational campaign that presents the benefit and overall gains and consultation to address the specific contentions and reach a compromise.

Modernisation options for the physical structures would not only cost more but would need some time for deliberation and forging agreement among engineering and operation personnel and water users on the overall system objective and operational objectives of Balanac RIS. Many of the options were improvement works whose necessity is readily observable by water users and field personnel. Although changing the type of headwork and additions of a downstream dam are not common, their merits are readily conceivable in the context of climate change. The high investment cost associated with this option would be an issue.

It is important to note that modernisation could be carried out in a progressive manner, starting with the relatively simple or costly improvements to the biggest ones. Some of the options could be modified and additions of new ones are possible. The most important is that the collectively shaped vision and general direction are translated into a modernisation master plan that will serve as a guide for succeeding improvements and can be handed down to the next generation of system officials and water users.

11 EVALUATION

11.1 Introduction

The main objective of the research was to formulate a strategy for developing a modernisation strategy for small national irrigation systems in the Philippines. The research methodology was framed with deliberately selected assessment and characterization procedures, which were adaptively modified and integrated to critically analyze the state of coherence among the fundamentals of irrigation system water delivery: design, operation, management and water supply; and to identify solutions for any inconsistency that hinders the achievement of the desired irrigation service. An analysis of the applicability of the integrated procedures and the relevance of the findings and insights obtained through them will pave the way for its evaluation as a plan of actions for developing a NIS modernisation plan. This chapter summarizes the key findings and conclusions for the objectives of the study. It discusses some recommendations to address the key challenges as well as directions for further development of the strategy. Finally it provides the conclusions of the study.

11.2 Key findings

11.2.1   Nature and impacts of rehabilitation in terms of closing the gap between the actual area irrigated and irrigation service area

With respect to the nature and impacts of rehbilitation on closing the gap between the actual area irrigated and the irrigation service area the following key findings can be formulated:

- despite considerable magnitude of rehabilitation of national irrigation systems carried out nationwide (122,200 ha yr^{-1}) and in the case study systems (780 ha yr^{-1} in or about 76% of FUSA of Balanac RIS and 330 ha yr^{-1} or 34% of FUSA of Sta. Maria RIS) during the 2000-2014 period, the gap between the actual area irrigated and irrigation service area remains. At the national level, about 70% of the developed service areas and O&M service areas were actually irrigated and the average irrigation intensities were 140%. In terms of the design service areas, about 60% were actually served and the annual irrigation intensity was 120%;

- rehabilitation had at best helped maintain an average actual irrigated area of around 870 ha of Balanace RIS and 840 ha of Sta. Maria RIS or 86% of their respective FUSA during the 2006–2015 period;

- canal lining was the most frequent and most invested in the case study systems, accounting for about 44% (Balanac RIS) and 51% (Sta. Maria RIS) of the total rehabilitation expenses. It was followed by headwork (38%) and road (17%) for Balanac RIS and Sta. Maria RIS, respectively. Canal structures accounted for less than 5% of the expenses for both systems;

- the underperformance of rehabilitation projects in terms of irrigating the whole service area is symptomatic of shortcomings of the rehabilitation approach to effectively address the root causes of mediocre irrigation service.

11.2.2   Efficacy of the physical structure and operations of irrigation systems to deliver the design rate, duration and frequency of irrigation water

With respect to the efficacy of the physical structure and operations of irrigation systems to deliver the design rate, duration and frequency of irrigation water the following key findings can be formulated:

- there was coherence among the design philosophy, objectives and flow control method in the original design of the case study systems. However, the shift to 'splitted' flow and 'proportional control' in Balanac RIS was inconsistent with the 'productive irrigation' objectives (that is, providing irrigation for optimum crop production). Similarly, the additions of open direct turnouts along Sta. Maria RIS main canals were incoherent with the 'adjustable flow' method of water allocation to second level canals and main canal and Lateral A offtakes as well as with the 'adjustable flow' method of water distribution through these main conveyance canals;

- there was much to be desired for the water delivery service (in terms of equity, reliability, flexibility and volume measurements), control and operation of canal structures, social order, strength of water users, budget and employees. These aspects had generally low performance ratings (0-2 in the scale of 0-4) in terms of the RAP internal indicators. The low values of the RAP internal indicators for water delivery and control and operation of flow structures for the two systems were mainly due to the lack of functional flow control structures and presence of direct, open turnouts along the main canals and other major conveyance canals;

- the capacity of the physical structures of the case study systems to perform their functions of conveyance, diversion, division, water level regulation, flow measurement, storage and discharge transfer and to satisfy present demands was either inherently limited or had decreased due to damages, dysfunctions, missing parts and deviations from preconditions for proper functioning through time. The most telling capacity issues were the reduced diversion capacity due to siltation and inherently limited storage capacities of the run-of-the river dams of both systems and poor division capacity of canal system of Balanac RIS due to lack of operational flow control structures;

- although originally designed to proportionally distribute any changes in canal discharge among the offtakes (hydraulic flexibility F = 1), most of the diversion points of Balanac RIS and Sta. Maria RIS would have high sensitivity (F > 1) due to the lack of functional flow control structures and highly variable incoming water supply;

- unrestrained direct offtakings and discharging into the canals, illegal flow obstructions and water supply fluctuation would be the main causes of perturbations;

- the present set and state of the dams and canal flow control structures would not support well the productive irrigation objective and equitable and wieldy water distribution.

11.2.3   To assess the validity of design values for dependable water supply, crop water requirement and percolation

With respect to the assessessment of the validity of design values for dependable water supply, crop water requirement, seepage and percolation the key findings are as follows:

- different parts of the service areas of Balanac RIS and Sta. Maria RIS could have different percolation rates. Four of the experiment sites each in Balanac RIS and

Sta. Maria RIS had higher percolation rates than the design values for the systems. Those in Balanac RIS were 5-14 times higher than the system design values of 2 mm day^{-1} while those in Sta. Maria RIS were about 2-5 times higher than the design rates of 1.3-1.4 mm day^{-1}. These higher percolation rates would result in gross underestimation of crop water requirement and irrigation requirement. There would be a need to determine the actual value in at least one representative site per turnout service area so that a more realistic estimate of irrigation requirement can be made;

- the low probability of exceedance (about 33% for SMMC and between 14-18% for MMC) of the assumed values for available water supply and the very low flows (about 200 l s^{-1} for SMMC and 60 l s^{-1} for MMC) associated with the conventional design probability of 80% when gleaned from their discharge rating curves were inconsistent with local knowledge on observed "usual flows", sample flow measurements and from the fact that the irrigation system had consistently irrigated an area of around 850 ha. The flumes would need to be recalibrated to revalidate the discharge rating curves;

- while there was good agreement between the programmed area (service area that can be irrigated based on an assumed flow situation for the season) and the actual area served, questions on trueness of the quesswork estimation of available water arised from the recognized fact that actual irrigated areas of Balanac RIS and Sta. Maria RIS did not differ much even though their water supply situations significantly differed;

- in-depth technical assessments of discharges of Balanac River and Sta. Maria River through flow gauging and flow modeling need to be pursued to better quantify the dependable water.

11.2.4   Identification of modernisation options for existing NIS service area

With respect to modernisation options he following key items can be formulated:
- the high demand for canal operation supervision required to achieved equitable and orderly water distribution in both systems can be reduced by closing the ungated direct offtakes along the main canals and providing operational, technically coherent and socially acceptable flow control structures;

- the different turnout service areas (TSA) were not consolidated enough as management units as most of them have several turnouts directly diverting water from the main conveyance canals. Such loose canal hierarchy impedes the formation of unified TSA for a more manageneable system operation. These turnouts would need to be consolidated in optimal number of farmditches based on hydraulic considerations;

- the downstream TSA of each system experienced water scarcity and submergence, especially during El Niño and La Niña years. A climate-proactive cropping calendar would need to be implemented to cope with the adverse climatic conditions, especially in Balanac RIS. Some of the strategy can be adopted and implemented included shifting of the cropping calendar to earlier dates, either from upstream farms or downstream farms, irrigation rotation and planting of less water-loving crops;

- drainage water and groundwater in the case study systems have been tapped to supplement the water supply from the dam. Development efforts to optimize the gains from such water would include the following: development of the drainage network and construction of additional water reuse structures (check gates); spring development in Sta. Maria RIS; and formulation of pump irrigation development guidelines (appropriate spacing, sustainable withdrawal rates); and a clear conjunctive use strategy and corresponding ISF adjustments;

- a storage-type dam would be a more logical headwork modernisation option in the context of climate change;

- the cost of frontline O&M activities in both systems were below the PhP 2,000 - 3,000 ha^{-1} standard planning budget and the PhP 1,900 ha^{-1} actual average expenses for O&M nationwide. Wages and salaries of field personnel accounted for about 70% of their respective WUA budgets. Only about 12% and 6% were spent for canal repair and maintenance in Sta. Maria RIS and Balanac RIS, respectively;

- multiple uses of water within the service areas such as those for aquaculture, livestock and vegetable farms and orchards were slowly emerging. They would need to be integrated in planning irrigation system modernization.

11.2.5   *Investigation of the potentials of shallow tubewell irrigation in the part of NIS service areas that cannot be fully irrigated with surface water as designed*

With respect to the potentials for shallow tubewell irrigation, the key findings are as follows:

- there is a good potential of groundwater for irrigation in downstream farms of Balanac RIS as evidenced by the actual usage of shallow tubewells by the farmers;
- difficult and deep well areas that were encountered during the well drilling in Sta. Maria RIS present a considerable challenge to tubewell irrigation development. A geo-resistivity survey would be useful for preliminary drilling explorations;
- the individual free-flowing wells used by local residents for domestic and aquaculture purposes indicate a good potential of spring water for supplemental irrigation. Development of the springs to augment the canal water supply would also be a viable alternative.

11.2.6   *Identification of possible options of integrating conjunctive use of groundwater and surface water in the formulation of modernisation plans for NIS*

With respect to the possibility of integrating conjunctive use in the modernisation plans, the key findings are as follows:

- the use of STW was a downstream farmers' initiative to supplement their water supply from the dam;
- STW units were operated as private irrigation scheme, independent of the canal network operation and system management of Balanac RIS;
- spring water runoff, together with drainage water from upstream farms is captured for irrigation by means of checkgates installed along natural drainage channels crisscrossing the service area of Sta. Maria RIS and are diverted to a TSA, which are either solely served by runoff water or served by both canal and runoff water;

- it would be best to maintain the independent operation of STW in Balanac RIS. The aspect where the integration of conjunctive use of surface water and groundwater would have good prospects would be on identifying the TSA that would need to use STW for each cropping season. Consequently, this will delineate the areas where water and canal operation can focus on.

11.3 Strategy to formulate a NIS modernisation plan

The key features of the proposed approach for developing NIS modernisation plans are the critical examination of the logical coherence among the design, operation and water supply with respect to the desired irrigation service and, consequently, the identification of technical and managerial solutions to address any inconsistencies. The proposed approach consisted of assessment and characterisation procedures that were adaptively modified, devised and pieced together to form a more comprehensive and relevant methodology for such analysis. The step-by-step methodology is as follows:

- Analysis of the nature and impacts of previous rehabilitation on the irrigation performance targets would give valuable insights on what intervention measures had been carried out, which did not worked and which have the potentials for improving irrigation service and closing the gap between the irrigation service area and actual irrigated areas. Categorization of the nature or type of rehabilitation and/or improvement works and system walkthrough would be essential in the analysis.

- Assessment of the efficacy of the physical structure and system operations to deliver the desired irrigation service. A combination of a logic design analysis and diagnostic tools of the MASSCOTE such as the RAP and assessment of system capacity, system sensitivity and perturbations, was formulated and used in the system diagnosis. The inherent constraints in the design of the system were identified through the former by examining the logical coherence among the design philosophy, overall system objectives, operational objectives, design configuration of the physical structures and flow control methods. The integrity and soundness of the existing canal structures, canal operation and ensuing irrigation service were assessed through the diagnostic tools of MASSCOTE.

The pragmatic or adaptive modifications done to address some pecularities in most NIS included (1) the use of the general discharge-head relation $Q = cH^{\alpha}$ and the diversion sensitivity or hydraulic flexibility values of commonly design configuration as conceptualized in Horst (1998) and Ankum (2001) in cases of lack of flow data and infeasibility of flow measurements; (2) review of the general design guidelines and history of the scheme development in cases of inavailability of original design documents; (3) use of the results of system walkthrough and field interview on qualitative ratings of the adequacy of irrigation water for each turnout service areas and occurrence of any tailwater, springs, pumped irrigation and creeks water in cases of lack of data on water balance parameters for the RAP external indicators; (4) and inclusion of headwork components in the combined diagnostic approach.

- Revalidation of design assumptions on percolation and water supply. The consistent gap between the design irrigation service area and actual area irrigated have been attributed to the broad application of standard design values and optimistic assumptions on percolation rates and dependable water supply. More reliable data must be used in the planning and design of irrigation modernisation projects. Field measurements of percolation rates in different turnout service areas and measurement and monitoring of the diverted water and river flows feeding the NIS have to be started to establish these crucial data. Modern flow estimation and simulation techniques would be useful in the assessment available water for a modernisation projects.

- Characterization of the surface water flow paths, system operation, irrigation service, unit service areas and demand for canal operations would facilitate a more holistic examination of the prevailing conditions of these components and, consequently, identification of options for improvements. In cases of lack of flow data, a sense of a realistic service area estimate for the prevailing water supply would also be apparent. The spatial analysis and analytical concepts and tools of the MASSCOTE approach in characterizing such system information would be useful. The pragmatic modifications that was adopted include: (1) characterisation of irrigation service for each turnout service area based on the RAP parameters for actual water delivery (volume measurement, flexibility, reliability and equity); (2) qualifying demand for canal operation for each turnout service area based on associated crop risk (occurence of water scarcity, flood and other water sources for supplemental irrigation), operational requirements (gated or ungated) and status (functional or dysfunctional) of flow diversion structures, and perturbations based on the total count of ungated direct offtakes upstream; and (3) Cost estimation based on front line irrigation services (operation, maintenance and management at the irrigation system level)

- Consolidation of options and visions for irrigation system modernisation. While technical and managerial solutions to the root causes of poor irrigation performance are identified through an objective methodology for irrigation modernisation study, it is crucial that these solutions would be a shared belief and easily relatable by system officials and water users. A workshop-consultation with system officials and turnout service area representatives would have to be carried out so proposed modernisation options can be discussed and a collective vision of the desired future scenarios of the irrigation system can be drawn. Sufficient information about the different modernisation options and their relevant implications would need to be provided to system officials and farmers so that they may be facilitated to make informed choices. Options for modernisation would need to be integrated and harmonized in a logical manner with the envisioned modernised irrigation system. The results of such consultation would be useful in fine-tuning and finalizing the draft of an irrigation system modernisation plan.

- The utility of the proposed methodology in the formulation of modernisation plan for NIS was demonstrated in the case of Balanac RIS and Sta. Maria RIS. The methodology would be applicable in other small- to medium-scale NIS and similar ungauged canal irrigation systems.

11.4 Challenges

This research has benefitted from and has successfully used and/or adaptively modified a number of existing methodologies, analytical tools, concepts and techniques to craft a methodology that would be relevant to the formulation of a modernisation plan in the case of government funded canal irrigation systems in the Philippines. However, some of the challenges remain, such as:

- the continued incommensurate actions to scientific determination of water supply availability with the articulated recognition of the paramount importance of water availability in prioritizing irrigation development and improvement projects illustrated by many ungauged NIS and water supply sources;

- the long-held view of irrigation as a social benefit rather than as an economic service does not stimulate innovative thinking or actions to modernise;

- management complacency, either a strongly held belief that irrigation systems are operated and managed in the best possible way or resigned acceptance to current situation and unconcerned with changing it;

- less critical use or adoption of generalized design values, assumptions and long existing technology conventionally used for irrigation system design;

- scarcity of data and information for planning for modernisation.

11.5 Recommendations

To further test the applicability of the identified modernisation options for NIS, the following reccomendations based on the findings of the research are forwarded:

- upon completion of the critical examination of the root causes for poor water delivery and identification of solutions, crafting of the comprehensive plan to improve it, planning for irrigation system modernisation need the support of and agreement by both irrigation agency officials and water users. Thus, it is important to have series of extensive consultations regarding the proposed modernisation options and general directions;

- a coherent link among the physical structures, system operation and water supply was central to the identified modernisation options and framing of the modernisation strategy for NIS. Nevertheless, economic considerations, rational planning and uniqueness of each NIS compel testing of feasibility and the actual functioning of the canal system with the identified or adjusted modernisation solutions through modelling or simulation of modernisation scenarios;

- flow gauging activity, estimation of water supply of ungauged rivers and determination of percolation rates are major undertakings of thesis topics by themselves and are crucial inputs to irrigation modernisation planning. These undertakings could be separately initiated to support planning for modernisation;

- with more distinct dry and wet seasons and high and low river flow regimes, storing water during times of plenty has increasingly become crucial to irrigated agriculture. Construction of reservoir or storage-type dams, though requiring major investments, would be needed to modernise the case study systems. Dams would have to be desilted and repaired in the short term modernisation plan. The technical feasibility of a second dam at Lower Balanac River would need to be studied. If found to be feasible, such a dam would imply cutting of canal dependence or delinking of lower canal operations with the upstream canals, hence operations can be simplified and water delivery can be improved;

- the cropping calendar and irrigation practices would need to be adjusted with changes in hydrology and climate. It would be necessary for Balanac RIS to adopt a coordinated, system-wide cultivation and irrigation schedule;

- water distribution and allocation in Balanace RIS would need to employ only one flow control method (proportional control or upstream control) and its corresponding flow control structures to avoid chaotic and unwieldy water distribution. The other structures would need to be removed or closed to avoid conflicting system management;

- the following canal-related improvement options are recommended: replacement of dysfunctional structures; repairs and desilting of canals; closing of ungated direct offtakes along the main canals; consolidation of turnouts along the main conveyance canals to main farm ditches and provision of flow measurement structures downstream of main intake structures and at distribution points;

- an official water distribution plan and canal operation instructions manual would have to be written and accessible to water users to promote transparency and accountability in system management.

11.6 Future activities

Some future activities arising from this research are as follows:

- river flow monitoring would need to be resumed to provide the necessary flow data inputs for programming areas to be irrigated and corresponding O&M plan, a more reliable estimation of water availability for modernisation projects and simulation of instantenous river flows for reservoir design and operation studies;

- in view of difficulty or sometimes practical infeasibility field measurement of sensitivity, simulation of the hydraulic behaviour of the canal network would be useful;

- an assessment of the sustainable yield of shallow aquifers and artesian wells in Balanac RIS and Sta. Maria RIS would need to be carried out to determine their potentials as supplemental sources and for conjunctive use;

- the strategy for planning for irrigation system modernisation developed in this research can be applied in some medium- to large-scale NIS to determine its utility in such NIS.

11.7 Conclusion

As demonstrated in this study, it can be stated that the set of procedures used in examining the different aspects of planning and operations of NIS with an end view of modernising the system provides a more comprehensive and applicable methodology for drawing up of a more relevant plan for NIS modernisation. The knowledge gained on the case study systems through this endeavor provides a sound basis for the design of appropriate modernisation solutions for the systems and other systems under similar conditions. The methodology developed in this study could serve as a blueprint for a modernisation program for NIS. It would also lead to more concerted system improvement projects and programs towards system modernisation.

Fundamental to the strenght of the modernization strategy is the strong emphasis on the coherent link among the system physical structure, operation and water supply during the examination of the state of irrigation systems and during the formulation of modernisation solutions. Nevertheless, external factors such as emerging multiple water uses, cultural norms, national laws, policy and development thrust of the government and pertaining to irrigation and water use could greatly influence directions and outcomes of modernisation programs. Indeed, irrigation system modernisation is not a fixed blueprint of course of actions but a process of upgrading irrigation system technology and management to provide better irrigation service.

12 REFERENCES

Abarabar (2014). *Soil characterization of the irrigated areas of Balanac River Irrigation Systems, Laguna, Philippines*. Unpublished undergraduate thesis: University of the Philippines Los Baños

Abdullah, K.B. (2006). Use of water and land for food security and environmental sustainability. *Irrigation and Drainage, 55,* 219–222.

Allen, R.G., Pereira, L.S., Raes, D. and Smith, M. (1998). *Crop evapotranspiration: Guidelines for computing crop water requirements*. FAO Irrigation and Drainage Paper 56. FAO, Rome Italy.

Ankum, P. (1997). *Selection of operation methods in canal irrigation delivery systems*. In: Proceedings ICID/FAO Workshop on Irrigation Scheduling. Rome, Italy. 12–13 Sept. 1995.

Ankum, P. (1999). International Misunderstanding in Irrigation Engineering. *Lowland Technology International, 1* (1), 47–48.

Ankum, P. (2001). *Flow control in irrigation systems*. Lecture notes. UNESCO-IHE, Delft, the Netherlands. 344 pages.

Araral Jr., E. (2005). Bureaucratic incentive, path dependence and foreign aid: an empirical institutional analysis of irrigation in the Philippines. *Policy Sciences, 38,* 131–157.

Asian Development Bank (ADB) (2008). *Republic of the Philippines: Preparing the Irrigation System Operation Efficiency Improvement Project*. Technical Assistance Report. Manila, Philippines.

Avila (2014). *Soil characterization of areas serviced by the Sta. Maria River Irrigation Systems, Laguna Philippines*. Unpublished undergraduate thesis: University of the Philippines Los Baños.

Bagadion, B. (1994). Joint management of the Libmanan-Cabusao pump irrigation system between farmers and the National Irrigation Administration in the Philippines. In: S.H. Johnson, D.L. Vermillion and J.A. Sagardoy (eds.), *Irrigation Management Transfer (p.* 313-330). International Irrigation Management Institute (IIMI) and FAO, Water Reports No. 5 (1995). FAO, Rome.

Bos, M.G. and Nugteren J. (1990). On irrigation efficiencies. 4th ed. International Institute for Land Reclamation and Improvement. Publication 19. Wageningen, the Netherlands. 24 pages.

Bos, M.G., Burton, M.A. and Molden, D.J. (2005). *Irrigation and Drainage Performance Assessment: Practical Guidelines.* CABI Publishing, Cambridge, USA.

Bruscoli, P, Bresci, E. and Preti, F. (2001). Diagnostic Analysis of an irrigation system in Andes Region. *Agricultural Engineering International: the CIGR EJournal of Scientific Research and Development* Vol. III, February.

Burt, C. (2001). Rapid *Appraisal Process (RAP) and benchmarking explanation and tools.* World Bank Irrigation Institution Window. World Bank, USA.

Burt, C.M. and Styles, S.W. (1999). Modern Water Control and Management Practices in Irrigation. Impact on Performance. FAO Water Report No. 19. Rome, Italy. 224 pages.

Burt, C.M. and Styles, S.W. (2004). Conceptualizing irrigation project modernization through benchmarking and the Rapid Appraisal Process. *Irrigation and Drainage, 53,* 145–154.

Burton, M.A., Kivumbi, D. and El-Askari, K. (1999). Opportunities and constraints to improving irrigation water management: Foci for research. *Agricultural Water Management, 40,* 37–44.

Burton, M. (2010). Irrigation management: principles and practices. In: Irrigation and Management Transfer and Organization Restructuring.

Chambers, R. and Carruthers I. (1986). *Rapid appraisal to improve canal irrigation performance: Experience and options.* IIMI Research Paper No.3. International Irrigation Management Institute, Colombo, Sri Lanka.

Chow, V.T. (1959). Open Channel Hydraulics. McGraw-Hill New York, USA.

CIRDUP (2007). *Comprehensive Irrigation Research and Development Umbrella Program: Completion Report.* University of the Philippines Los Baños Foundation, Inc. and Department of Agriculture. Quezon City, Philippines.

Clemmens, A.J., Holly, F.M. and Schuurmans, W. (1993). Description and Evaluation of Program: DUFLOW. *J. Irrig. and Drain. Engrg, 119* (4), 724–734.

Clemmens, A.J. (2006). Improving irrigated agriculture performance through an understanding of the water delivery process. *Irrigation and Drainage, 55,* 223–234.

Clemmens, A.J., Bautista, E., Wahlin, B.T. and Strand, R.J. (2005). Simulation of Automatic Canal Control Systems. *J. Irrig. and Drain., 131* (4), 324–335.

David, W.P. (1997). Accelerating Transformation to Irrigated Agriculture: An imperative for agricultural modernization. Working paper on Irrigation, Agricultural Commission (AGRICOM). Congress of the Philippines, Manila.

David, W.P. (2003). *Averting the Water Crises in Agriculture: Policy and Program Framework for Irrigation Development in the Philippines.* University of the Philippines Press, Diliman, Quezon City, Philippines.

David, W.P. (2006). *Agriculture and Fisheries Modernization Act (AFMA) Review: Irrigation and SAFDZ Components.* Report to the Congressional Oversight Committee on Agriculture and Fisheries Modernization (COCAFM). Congress of the Philippines. Manila.

David, W.P. (2007). *Advanced Water Resources Planning.* AENG 240 Course Manual. University of the Philippines Los Baños. Laguna, Philippines.

David, W.P. (2008). Irrigation (Chapter 6): In *Modernizing Philippine Agriculture and Fisheries: The AFMA Implementation Experience.* Dy, R.T., Gonzales, L.A., Bonifacio, M.F., David, W.P., De Vera III, J.P.E., Lantican, F.A., Llanto, G.M., Martinez, L.O., Tan, E.E. University of Asia and the Pacific. Manila, Philippines.

David, W.P. (2009). Impact of AFMA on Irrigation and Irrigated Agriculture. *The Philippine Agricultural Scientist, 91* (3) 315–328.

David, W.P. and Delos Reyes, M.L. (2003). *Program Framework for Irrigation Development.* Paper Presented at the Workshop on DAR-DA-DSWD-DOST Convergence Program. Sulu Hotel, Quezon City. March 6, 2003.

Dedrick, A.R., Bautista, E., Clyma, W., Levine, D.B., Rish, S.A. and Clemmens, A.J. (2000). Diagnostic analysis of the Maricopa-Stanfield Irrigation and Drainage District area. *Irrigation and Drainage Systems, 14,* 41–67.

Depeweg, H. and Paudel K.P. (2003). Sediment transport problems in Nepal evaluated by the SETRIC model. *Irrigation and Drainage, 52,* 247–260.

Droogers, P. and Allen, R.G. (2002). Estimating reference evapotranspiration under inaccurate data conditions. *Irrigation and Drainage Systems, 16,* 33–42.

Dy, R. (1990). *Economic Evaluation of Communal Irrigation Systems*. Report prepared for the World Bank. Pasig, City Philippines. University of Asia and the Pacific.

Ertsen, M.W. (2007). The development of irrigation design schools or how history structures human action. *Irrigation and Drainage, 56*, 1–19.

Ertsen, M.W. (2009). From central control to service delivery? Reflections on irrigation management and expertise. *Irrigation and Drainage, 58*, S87–S103.

Facon, T., Renault, D., Rao, P.S. and Wahaj, R. (2008). High-yielding capacity building in irrigation system management: targeting managers and operators. *Irrigation and Drainage, 57*, 288–299.

Ferguson, C. (1987). *Returns to Irrigation Intensification in Philippine Gravity Systems*. Unpublished Ph.D. Dissertation, Cornell University, Ithaca, New York.

Food and Agriculture Organisation of the United Nations (FAO) (2007). The relationship between water, agriculture, food security and poverty. FAO, Rome, Italy.

Ghumman, A.R., Khan, M.Z., Khan, A.H. and Munir, S. (2010). Assessment of operational strategies for logical and optimal use of irrigation water in a downstream control system. *Irrigation and Drainage, 59*, 117–128.

Ghumman, A.R., Khan, Z. and Turral, H. (2009). Study of feasibility of night-closure irrigation canals for water saving. *Agricultural Water Management, 96* (3) 457-464.

Gorantiwar, S.D. and Smout, I.K. (2005). Performance assessment of irrigation water management of heterogeneous irrigation schemes: 1. A framework for evaluation. *Irrigation and Drainage Systems, 19*, 1–36.

Hales, A.L. and Burton, M.A. (2000). Using the IRMOS model for diagnostic analysis and performance enhancement of the Rio Cobre Irrigation Scheme, Jamaica. *Agricultural Water Management, 45*, 185-202.

Holly, F.M. and Parrish, J.B. (1993). Description and Evaluation of Program: CARIMA. *J. Irrig. and Drain. Engrg, 119* (4), 703–713.

Horst, L. (1998). *The Dilemmas of Water Division: Considerations and Criteria for Irrigation Systems Design*. International Water Management Institute. Colombo, Sri Lanka.

Hussain, I. (2007)a. Poverty-reducing impacts of irrigation: Evidence and lessons. *Irrigation and Drainage, 56*, 147–164.

Hussain, I. (2007)b. Pro-poor intervention strategies in irrigated agriculture in Asia: Issues, lessons, options and guidelines. *Irrig. and Drain., 56*, 119–126.

International Programme for Technology and Research in Irrigation and Drainage (IPTRID) (1999). *Poverty reduction and irrigated agriculture..* Issue Paper No. 1. FAO, Rome, Italy.

International Programme for Technology and Research in Irrigation and Drainage (IPTRID) (2003). *The irrigation challenge: Increasing irrigation contribution to food security through higher water productivity form canal irrigation.* Issue Paper No. 4. FAO, Rome, Italy.

Khan, M.Z. and Ghumman, A.R. (2008), Hydrodynamic modelling for water-saving strategies in irrigation canals. *Irrigation and Drainage, 57*, 400–410.

Kouchakzadeh, S. and Montazar, A. (2005). Hydraulic sensitivity indicators for canal operation assessment. *Irrig. and Drain., 54*, 443–454.

Kraatz, D.B. (1977). *Irrigation canal lining.* FAO Land and Water Development Series No. 1, 1977, 18–50. FAO, Rome. Italy.

Kumar, P., Mishra, A., Raghuwanshi, N.S. and Singh, R. (2002). Application of unsteady flow hydraulic-model to a large and complex irrigation system. *Agricultural Water Management, 54*, 49-66.

Lipton, M. (2007). Farm water and rural poverty reduction in developing Asia. *Irrigation and Drainage, 56*, 127–146.

Litrico, X., Fromion, V., Baume, J., Arranja, C. and Rijo, M. (2005). Experimental validation of a methodology to control irrigation canals based on Saint-Venant equations. *Control Engineering Practice, 13*, 1425–1437.

Malano, H.M. and Hofwegen P.J.M. (2006). *Management of Irrigation and Drainage Systems - A Service Approach.* UNESCO-IHE Monograph 3. Taylor and Francis/Balkema, Leiden, The Netherlands.

Maleza, M.C.E. and Nishimura, Y. (2007). Participatory Processes and Outcomes: The case of national irrigation management in Bohol, Philippines. *Irrigation and Drainage, 56*, 21–28.

Malik, R.P.S, Prathapar, S.A; Marwah, M. (2014). *Revitalizing Canal Irrigation: Towards Improving Cost Recovery*. IWMI Working Paper 160. International Water Management Institute. Colombo, Sri Lanka.

McLoughin P.F.M. (1988). O&M spending levels in third world irrigation systems: exploring economic alternatives. *Water Resources Bulletin*. American Water Resources Association, 24 (3), 599-607.

Meijer, T.K.E. (1992). Three Pitfalls in Irrigation Design. In: G. Diemer and J. Slabbers (eds.), *Irrigators and Engineers*. Thesis Publisher. Amsterdam, The Netherlands.

Merkley, G.P. (2006). RootCanal: user's guide and technical reference. Biological and Irrig. Engrg. Dept., Utah State University, Logan, Utah, USA.

Merkley, G.P. and Rogers, D.C. (1993). Description and Evaluation of Program CANAL. *J. Irrig. and Drain. Engrg., 119* (4), 714–723.

Mishra, A., Anand, A., Singh, R. and Raghuwanshi, RS. (2001). Hydraulic Modelling of Kangsabati Main Canal for Performance Assessment. *J. Irrig. and Drain. Engrg., 127* (1), 27–34.

Molden, D. (1997). *Accounting for Water Use and Productivity*. SWIM Paper 1. International Irrigation Management Institute. Colombo, Sri Lanka.

Molden, D., Burton, M. and Bos, M.G. (2007). Performance assessment, irrigation service delivery and poverty reduction: Benefits of improved system management. *Irrig. and Drain., 56,* 301–320.

Murray-Rust, D.H. and Snellen, W.B. (1993). Irrigation System Performance Assessment and Diagnosis. Joint IIMI/ILRI/IHEE Publication. International Irrigation Management Institute, Colombo, Sri Lanka.

National Irrigation Administration (NIA) (1990). *A comprehensive history of irrigation in the Philippines*. Quezon City, Philippines.

National Water Resources Council (NWRC) (1982). *Rapid Assessment of Water Supply Sources*. Vol. 1-73. Quezon City, Philippines.

National Water Resources Council (NWRC) (1980). Philippine Water Resources Summary Data. Volume 1 - Streamflow and Lake or River Stage. Ending December 31, 1970. Report No. 9. January 1980. Quezon City, Philippines.

Oad, R., McCornick, P.G. and Clyma, W. (1988). *Methodologies for interdisciplinary diagnosis of irrigations systems: Review and analysis of the methodologies used for irrigation system Diagnosis under the Water Management Synthesis II Project*. WMS Report 93. Colorado State University, Fort Collins, USA.

Ojo, O.I. and Otieno, F.A. (2010). Irrigation canal simulation models and its application to large scale irrigation schemes in South Africa: A review. *OIDA International Journal of Sustainable Development, 1* (2) 55–60.

Okada, H., Styles, S.W. and Grismer, M.E. (2008). Application of the Analytic Hierarchy Process to irrigation project improvement, Part I: Impacts of irrigation project internal processes on crop yields. *Agricultural Water Management, 95,* 199–204.

Philippine Atmospheric, Geophysical and Astronomical Services Administration (PAGASA) (2015). *Climate map of the Philippines*. Retrieved from: http://www.pagasa.dost.gov.ph/ index.php/climate-of-the-philippines.

Philippine Atmospheric, Geophysical and Astronomical Services Administration (PAGASA) (2015). Monthly rainfall, temperature, relative humidity, evaporation data for the National Agro-meteorological Station (NAS) - University of the Philippines Los Baños (UPLB). Record files of the Agro-meteorological and Farm Structures Division (AFSD), UPLB, Laguna, Philippines.

Perry, C. (2007). Efficient irrigation; inefficient communication; flawed recommendations. *Irrigation and Drainage, 56,* 367–378.

Philippine Food Security Information System (PhilFSIS) (2014). *Tropical cyclone frequency, by item and year*. Philippine Statistics Authority (PSA). Retrieved from: http://philfsis.psa.gov.ph/index.php/id/15/matrix/J30FSTII.

Philippine Statistics Authority (PSA) (2010). *Philippine Standard Geographic Code (PSGC) Interactive*. Retrieved from: http://www.nscb.gov.ph/activestats/psgc.

Philippine Statistics Authority (PSA) (2014). *2014 Compendium of Philippine Environment Statistics*. PSA, Makati City, Philippines.

Playan, E. and Mateos, L. (2006). Modernization and optimization of irrigation systems to increase water productivity. *Agricultural Water Management, 80,* 100–116.

Plusquellec, H. (2002). *How Design, Management and Policy Affect the Performance of Irrigation Projects*. FAO Regional Office for Asia and the Pacific, Bangkok, Thailand.

Plusquellec, H., Burt, C. and Wolter, H.W. (1994). *Modern Water Control in Irrigation: Concepts, Issues, and Applications*. World Bank Technical Paper 246. Irrigation and Drainage Series. The World Bank. Washington D.C., USA.

Raby, N. (2000). Participatory irrigation management in the Philippines: National Irrigation Systems. In: Groenfeldt D. and Svendsen M. (eds.), *Case Studies in Participatory Management*. World Bank Institute, Washington D.C., USA.

Ravago, M.V. and Balisacan, A.M. (2015). Current Structure and Future Challenges of the Agriculture Sector. In *The future of Philippine Agriculture: scenarios, policies and investments under climate change*. Rosegrant, M., Sombilla, M. & Balisacan, A.M. Retrieved from : *http://www.econ.upd.edu.ph/dp/index.php/dp/article/view/1481/963*

Renault, D. (1999). Offtake Sensitivity, Operation Effectiveness and Performance of Irrigation System. *J. Irrig. and Drain., 125 (3)*, 137–147.

Renault, D. (2000). Aggregated hydraulic sensitivity indicators for irrigation system behaviour. *Agricultural Water Management, 43*, 151–171.

Renault, D. (2001). Re-engineering irrigation management and system operations. *Agricultural Water Management, 47,* 211–226.

Renault, D. (2008). *Sensitivity of Analysis of irrigation structures: Technical Briefs*. FAO Water Development and Management Unit, Rome, Italy.

Renault D, Godaliyadda GGA. 1999. *Generic typology for irrigation systems operation*. Research Report29. International Water Management Institute. Colombo, Sri Lanka

Renault, D., Facon, T. and Wahaj, R. (2007). *Modernizing irrigation management – the MASSCOTE approach*. FAO, Rome, Italy.

Rice, E.B. (1997). *Paddy Irrigation and Water Management in Southeast Asia. A World Bank Operations Evaluation Study*. The International Bank for Reconstruction and Development, The World Bank. Washington D.C., USA

Rogers, D.C. and Merkley, G.P. (1993). Description and Evaluation of Program USM. *J. Irrig. and Drain. Engrg., 119* (4) 693–702.

Schultz, B. and De Wrachien, D. (2002). Irrigation and drainage systems research and development in the 21st century. *Irrigation and Drainage, 51,* 311–327.

Schultz, B., Thatte, C.D. and Labhsetwar, V.K. (2005). Irrigation and Drainage. Main Contributors to Global Food Production. *Irrigation and Drainage, 54,* 263–278.

Schultz, B., Tardieu, H. and Vidal, A. (2009). Role of water management for global food production and poverty alleviation. *Irrigation and Drainage, 58,* S3–S21.

Schuurmans, W. (1993). Description and Evaluation of Program MODIS. *J. Irrig. and Drain. Engrg., 119* (4) 735–742.

Shahrokhnia, M.A. and Javan, M. (2005). Performance assessment of Doroodzan irrigation network by steady state hydraulic modeling. *Irrigation and Drainage Systems, 19,* 189-2006.

Skutsch, J. and Rydzewski, J.R. (2001). *Review of research and development needs in irrigation and drainage.* IPTRID, FAO, Rome, Italy.

Small, L.E. and Svendsen, M. (1992). A Framework for Assessing Irrigation Performance. IFPRI Working Papers on Irrigation Performance No.   1. International Food Policy Research Institute, Washington, DC, USA.

Tariq, J.A. and Latif, M. (2010). Improving Operational Performance of Farmers Managed Distributary Canal using SIC Hydraulic Model. *Water Resources Management, 24* (12) 3085–3099.

Turral, H., Svendsen, M. and Faures, J.M. (2010). Investing in irrigation: Reviewing the past and looking to the future. *Agricultural Water Management, 97,* 551–560.

UNCED (1992). *Agenda 21: The United Nations Programme of Action from Rio.* United Nations Conference on Environment and Development (UNCED), Rio de Janerio, Brazil, 3–14 June 1992.

USAID (1996). *Using Rapid Appraisal Methods.* Performance Monitoring and Evaluation, United States Agency for International Development (USAID) Center for Development Information and Evaluation. Retrieved from: http://pdf.usaid.gov/pdf_docs/ PNABY209.pdf

USAID (2010). *Using Rapid Appraisal Methods.* Performance Monitoring and Evaluation, United States Agency for International Development (USAID) Center for Development Information and Evaluation. 2nd ed. Retrieved from: http://www.usaid.gov/policy/ evalweb/documents/TIPS-UsingRapidAppraisalMethods.pdf

Wijayaratna, C.M. and Vermillion, D.L. (1994). *Irrigation management turnover in the Philippines: strategy of the National Irrigation Administration.* Report No. 4. Short Report Series. International Irrigation Management Institute. Colombo, Sri Lanka.

Vermillion, D.L. (2005). Irrigation sector reform in Asia: from "participation with patronage" to "empowerment with accountability". In: *Asian Irrrigation in Transition: Responding to Challenges*. Shivakoti, G.P, Vermillion, D.L., Lam, W.F., Ostrom, E., Pradhan, U., Yoder, R. Sage Publications India Pvt Ltd. New Delhi.

World Bank (1992). Philippines: Irrigated Agriculture Sector Review. Report No. 9848-PH. The World Bank. Washington, D.C., USA.

World Bank (1996). Irrigation O&M and System Performance in Southeast Asia: An OED Impact Study. Report No. 15824-PH. Operations Evaluation Department, The World Bank. Washington, D.C., USA.

ANNEXES

Annex A. List of symbols

Symbol	Description	Unit
$A(x, t)$	wetted cross-sectional area	m^2
d_1	depth of water at the beginning of measurements	m
d_2	depth of water after 24 hr	m
ET_c	crop evapotranspiration	m^3
ET_o	reference evapotranspiration	mm
F	hydraulic flexibility indicator	--
g	gravitational acceleration	ms^{-2}
H	head exercised on a structure	M
k_c	crop coefficient	--
k_M	flow factor related to the canal roughness or the friction factor $1/n$	--
L	length of the isolated canal section	M
m_i	coefficient of water depth variation within a backwater curve at any location i along a reach and at downstream cross regulator	--
N	Manning coefficient	$sm^{-1/3}$
P	average wetted perimeter	M
$Q(x, t)$	discharge across section A	m^3s^{-1}
q_i	discharge through the offtake i	m^3s^{-1}
R	hydraulic radius or the quotient of water cross-sectional area and wetted perimeter	M
S	energy gradient or the bed slope for canals with normal flow	--
S_b	bed slope	--
$S_{conveyance}$	sensitivity indicator for conveyance	m^{-1}

$S_f(x, t)$	friction slope	
S_G	sensitivity indicator for a downstream cross-regulator	m^{-1}
$S_i(i)$	sensitivity of the delivery for offtake i	m^{-1}
$S_{offtake}$	offtake sensitivity indicator	m^{-1}
S_{RC}	reach sensitivity indicator for conveyance	m^{-1}
S_{RD}	reach sensitivity indicator for delivery	m^{-1}
$S_{regulator}$	water level control (cross-regulator) sensitivity indicator	m^{-1}
S_{RH}	reach sensitivity indicator for water depth	m^{-1}
V	flow velocity	ms^{-1}
W	average width of water surface of the ponded canal section	M
$Y(x, t)$	water depth	M
A	exponent in hydraulic equation for flow	--
ΔH	variation in water level in a parent canal	M
$\Delta H_{permissible}$	permissible variation in water level in parent canal	M
Δq	variation of discharge through an offtake	m^3s^{-1}
$\Delta Q/Q$	relative discharge variation in the parent canal	--
$\Delta q/q$	relative variation in discharge through an offtake	--
ΔQ_{del}	variation of discharge delivered within a canal reach	m^3s
ΔQ_{in}	variation of discharge entering a canal reach	m^3s
ΔQ_{out}	variation of discharge leaving a canal reach	m^3s

Annex B. Acronyms

AFMA	Agriculture and Fisheries Modernisation Act
BSWM	Bureau of Soil and Water Management
DA	Department of Agriculture
DUFLOW	Dutch Flow
FAO	Food and Agricultural Organization of the United Nations
Ha	Hectares
ICID	International Commission on Irrigation and Drainage
ICSM	Irrigation Canal Simulation Model
IPTRID	International Programme for Technology and Research in Irrigation and Drainage
MASSCOTE	Mapping System and Services for Canal Operation Techniques
NIA	National International Irrigation Administration
NIS	National Irrigation System
O&M	Operation and maintenance
RAP	Rapid Appraisal Procedure
SIC	Simulation of Irrigation Canal
STW	Shallow tubewell
UPLB	University of the Philippines Los Baños

Annex C. Major rehabilitation and improvement projects

Foreign assisted projects	Brief description	System
1. National Irrigation System Improvement Project 1 (NISIP 1) May 1977 - Dec 1985 US$ 76 million (PhP 888.5 million)	Rehabilitated and upgraded 21 NIS (expanding some of them), constructed 3 new systems to serve 44,400 ha; strengthen NIAs O&M capability by constructing buildings, service facilities, equipment and vehicles and spare parts. reducing risk of schistosomiasis; feasibility of land consolidation	• Region 1: Ilocos Norte Irrigation Systems (Laoag-Vintar, Bolo, Pasuguin, Dingras, Cura, Bonga 1-3); Amburayan RIS; Masalip RIS; Ilocos Sur Irrigation System (Sta. Maria-Burgos, Sta. Lucia-Candon, Tagudin, Gaco) • Region 2: Banurbur Creek Irrigation System; Pamplona Pump Irrigation System • Region 8: Binahaan-Tibak-Guinarona-MP RIS; Hindang-Hilongos Irrigation System; Das-ay RIS; Bito-Daguitan, BIG Irrigation System
2. NISIP 2 Jun 1978 - Jul 1987 US$ 80.7 million (PhP 996.8 million)	Rehabilitated, upgraded and expanded 26 NIS to serve 76,300 ha. Provision of buildings, service facilities, equipment and vehicles and spare parts needed to strengthen NIA O&M capability; introduced agri input-output monitoring system; reduction of schistosomiasis and snail population, 4 FS and 1 DE of 4 irrigation projects	26 irrigation systems in Regions 4, 5, 6, 9 and 12 including feasibility studies for four irrigation projects (Balog-Balog Multipurpose, Lower Agno Multipurpose, Casecnan Transbasin Diversion and Jalaur Multipurpose). (The rest of the specific systems covered were not found in PCR)

Modernisation strategy for the national irrigation systems in the Philippines

Annex C. *Continued*

Foreign assisted projects	Brief description	System
3. Second Laguna De Bay Irrigation Project 1981-1992. U$40.2 million	• objectives: provide irrigation and infrastructure facilities required to increase agricultural production and farm incomes; create employment opportunities; and improve social conditions of the rural population in the project area. • components: augmentation of dry season irrigation water supply for the Cavite Friar Lands by pumping water from Laguna Lake through a pumping station and appurtenant structures; construction of 30-km feeder canal in teh Cavite Friar Lands; construction of a Balanac diversion dam; rehabilitation and improvement of the existing irrigation system covering Pagsanjan, East Bay area and Cavite Friar Lands; establishment of a vegetable production training center and relevant support services; consulting services for detailed engineering design and technical specifications for the pumping stations and pumps intallation and for planning and design for the vegetable component.	Cavite Friar Lands Irrigation System, Balanac River Irrigation System, Mayor River Irrigation System

Annex C. *Continued*

4.	IOSP 1 Jun 1988 - Jun 1993 US$ 69.1 million (PhP 1,714 million)	• objective: Strengthen institutional and technical capability of NIA and IAs in order to improve and maintain the efficiency of existing NIS structures; and to improve operating performance of the NIS through minor rehab works and increases in annual funding required to improve the level of O&M service • the first 3-yr phase of NIA's 9-yr irrigation O&M improvement program that would cover all NIS service area of about 500,000 ha • Components: rehabilitation works in NIS covering about 600,000 ha; expanded O&M program in all NIS; strengthening NIA capability on ISF collection process, cadastral surveys, procurement, implementation of irrigation management information system, and preparation of O&M manuals; and development program on organizing and training farmers in 114, 000 ha to assume responsibilities for O&M of laterals and sublaterals.	The list of NIS covered is found in Project Completion Report (PCR)
5.	Irrigation Systems Improvement Project 1 (ISIP 1) Mar 1991 - Dec 1996 US$ 36.3 million	Components: • rehabilitation and improvement of irrigation systems involving diversion headworks, canal networks, control structures, drainage, on-farm facilities and service roads (US$ 52.2 million) • construction of erosion control measures in catchment areas • institutional development • provision of construction and maintenance equipment and service vehicle • consulting services for supervision; support for schistosomiasis; M&E.	5 ADB-financed irrigation systems in Mindanao: Banga, Marbel, Pulangui, Saug, Simulao

Modernisation strategy for the national irrigation systems in the Philippines

Annex C. *Continued*

Foreign assisted projects	Brief description	System
6. IOSP II Oct 1993 - Dec 2000 US$ 51.3 million (US$ 1 = PhP 50)	• objective: Achieve sustainable improvement in the operational efficiency of NIS • components: (a) improvement of 18 selected NIS; urgent repair of 22 structures in 14 NIS; construction of three pilot sediment exclusion structures and a few improved pilot water control structures on a selected lateral; and erosion prevention measures in critical areas within and in the vicinity of existing NIS; (b) support to sustain the improved system-level O&M achieved under IOSP I; (c) institutional development through support to IAs and NIA; and (d) strengthening of agricultural support services.	Bonga 1 - 3, Cura, IAAPIS, Baggao, MRIIS 1-3, Agos, Matogdon, Barit, Sta. Maria, Pongso, Maranding, Malasila, Cantingas
7. ISIP 2 Feb 1996 - Apr 2006 US$ 34.78 million	Component 1: Physical infrastructure development (US$ 29.58 million). (a) remodeling of the existing irrigation service area of 12,649 ha in 9 systems; (b) improvement of existing diversions and distribution networks; (c) improvements to drainage infrastructure; (d) reconstruction of the Guinarona diversion, emergency repairs to the Lower Binahaan supply works, rehabilitation of the Daguitan headworks, and construction of new Marabong dam; (f) improvement to existing roads and construction of new roads; (g) institution of measures to achieve and sustain high standards for O&M, improve in ISF collection and clear delegation of O&M responsibilities. Component 2: Institutional development (US$ 3.7 million) Component 3: Agricultural improvement (US$ 0.76 million) Component 4: Environmental and social improvement and monitoring (US$ 0.74 million)	Leyte: Bao, Bito, Mainit, Binahaan North, Binahaan South, Lower Binahaan, Tibak, Daguitan, Guinarona, Marabong

Annex C. *Continued*

Foreign assisted projects	Brief description	System
8. Water Resources Development Program Mar 1997- Jun 2005 US$ 58.5 million (PhP 2,663 million)	Components: • improved Water Resources Planning and Management (US$ 4.5 million): preparation of a national water resource plan; improvement of national data collection networks; establishment of a national water information network; and strengthening the NWRB • improved Watershed Management (US$ 15 million): formulation of a national river basin management strategy and investment and institutional strengthening program; investments for improving river basin management in a number of identified priority river basins; and capability-building activities for staff of key national government agencies • improvement of National Irrigation Systems (NIS) (US$ 61.2 million): improvement of 14 irrigation systems, repair of eight major structures in other NIS, strengthening of the Masiway dam, and construction of sediment exclusion structures in another five NIS. • institutional Strengthening of NIA and IAs (US$ 3 million): provision of staff training, consultancies, and incremental operating costs to facilitate progressive turnover of systems • environment improvement (US$ 1.4 million): control of schistosomiasis in 3 NIS and control erosion in 31 NIS, establishment of an Environment Unit within the NIA and provision for training and consultancy to the Unit	The list of NIS covered is found in Staff Appraisal Report

Modernisation strategy for the national irrigation systems in the Philippines

Annex C. *Continued*

Foreign assisted projects	Brief description	System
9. Southern Philippines Irrigation Sector Project Oct 1999-30 Jun 2011 US$ 80.94 million	Components: participation and irrigation transfer component, physical infrastructure component	ARMM and Agusan Del Sur: Calagayon CIS, Can-asujan SRIS, Gibong Right Bank, Aclan Amontay CIS, Dauin SRIS
10. Irrigation System Operation Efficiency Improvement Project Nov 2008- Aug 2011 US$ 1.25 million	Improvement of irrigation services and institutional arrangements that increase the role of water users in system O&M; upgrading rural infrastructure, including irrigation systems, roads, and postharvest facilities of selected irrigation systems in Mindanao and the Visayas	Preparation/assessment/formu lation study: Mindanao: Saug River Multipurpose Project (NIS); Saug RIS (NIS); Libuganon RIS (NIS); Balagunan CIS; Sta. Lucia CIS; Upper Tuganay CIS Visayas: Aklan RIS (NIS); Sibalom-San Jose RIS (NIS)

Annex C. *Continued*

Foreign assisted projects	Brief description	System
11. Participatory Irrigation Development Project (PIDP) Phase 1: June 2009 - Mar 2015 Phase 1: Dec 2014 - Dec 2019 Phase 1: Sep 2018 - June 2024 US$ 413.59 million	Components: • irrigation sector restructuring and reform. Provide support to the implementation of NIA's Rationalization Plan, incorporating the program of severance payments and the corresponding institutional strengthening activities; establish more sustainable financial and institutional mechanisms for improved participatory O&M and routine rehabilitation • irrigation infrastructure development. Improve the delivery of irrigation services in about 58 NIS through rehabilitation and modernisation of irrigation systems in order to provide more reliable and flexible services • project management and coordination. Provide the support for an efficient coordination, implementation and management of the project, including strengthening the financial management and procurement functions and the establishment and operation of the results M&E system for the project	The list of NIS covered is found in project profile documents and basic information flyer
12. National Irrigation Sector Rehabilitation and Improvement Project (NISRIP) PhP4.07 billion January 2013 - December 2016	• objective: restore 11,501ha and rehabilitate 24,169 ha to increase rice productivity and establish sustainable O&M through rehabilitation of irrigation facilities, strengthening of irrigators associations (IA), and provision of IA facilities, equipment and agricultural support services.	The list of NIS covered is found in project profile documents and basic information flyer

Annex C. *Continued*

Locally funded projects	
1. FY2006 Irrigation Development Programs P500M PUMP PRIMING PROJECTS REPAIR/REHABILITATION OF NIS/CIS Under NDC 1 Funds As of December 31, 2009	SUMMARY BY SONA PHILIPPINES: NLAQ - CAR, I, ARIIP, II, MARIIS, III, UPRIIS MLUB - III, IV CP - IV, V, VI, VII, VIII MSR - IX, X, XI, XII, XIII By region, province and irrigation system/project: CAR, I, ARIIP, II, MRIIS, III (NLAQ), UPRIIS, III (MLUB), IV (MLUB), IV (CP), V, VI, VII, VIII, IX, X, XI, XII, XIII

Annex C. *Continued*

Locally funded projects	
2. FY2006 Irrigation Development Programs P500M PUMP PRIMING PROJECTS REPAIR/REHABILITATION OF NIS/CIS Under NDC 2 Funds As of December 31, 2009	SUMMARY BY SONA PHILIPPINES: NLAQ - CAR, I, II, MARIIS, III, UPRIIS MLUB - III, IV CP - IV, V, VI, VII, VIII MSR - IX, X, XI, XII, XIII By region, province and irrigation system/project: CAR, I, II, MRIIS, III(NLAQ), UPRIIS, III(MLUB), (MLUB), IV(CP), V, VI, VII, VIII, IX, X, XI, XIII
3. Irrigation Development Programs RESTORATION/REPAIR/REHABILITATION OF NIA-ASSISTED IRRIGATION SYSTEMS (P1.50B) CY 2007 (P3.126B) As of December 31, 2009	SUMMARY BY SONA PHILIPPINES NLAQ - CAR, I, ARIIP, II, MRIIS, III, UPRIIS MLUB - III, IV-A, IV-B CP - IV-B, V, VI, VII, VIII MSR - IX, X, XI, XII, XIII By region, province and irrigation system/project CAR, R1, ARIIP, R2, MRIIS, R3 (NLAQ), UPRIIS, R3 (MLUB), R4 (MLUB), R4 (CP), R5, R6, R7, R8, R9, R10, R11, R12, R13

Modernisation strategy for the national irrigation systems in the Philippines

Annex C. *Continued*

Locally funded projects	
4. Irrigation Development Programs RESTORATION/REPAIR/REHABILITATION OF NIA- ASSISTED IRRIGATION SYSTEMS (CY 2007 Augmentation Fund P1.5B) + (CY 2008 P1.65B) As of Apr 8, 2011	SUMMARY BY SONA PHILIPPINES: LUZON - CAR, I, ARIIP, II, MARIIS, III, UPRIIS, IV, V, VI VISAYAS - VII, VIII MINDANAO - IX, X, XI, XII XIII By region, province and irrigation system/project: CAR, I, ARIIP, II, MARIIS, III (NLAQ), III (MLUB), UPRIIS, IV-A (MLUB), IV-B (MLUB), IV-B (CP), V, VI, VII, VII, IX, X, XI, XII, XIII
5. Irrigation Development Programs RESTORATION/REPAIR/REHABILITATION OF NIA- ASSISTED IRRIGATION SYSTEMS NDC 6 & P6.524B (CY 2010 Allocation) As of April 30, 2011	SUMMARY BY SONA PHILIPPINES: NDC-6 P6.524B OTHERS LUZON - CAR, I, ARIIP, II, MARIIS, III, UPRIIS, IV-A, IV-B, V VISAYAS - VI, VII, VIII MINDANAO - IX, X, XI, XII, XIII By region, province and irrigation system/project CAR, I, ARIIP, II, MARIIS, III (NLAQ), III (MLUB), UPRIIS, IV, V, VI, VII, VIII, IX, X, XI, XII, XIII

Annex C. *Continued*

Locally funded projects	
6. RESTORATION/REHABILITATION OF NIA-ASSISTED IRRIGATION SYSTEMS NDC 5 Fund (P2.0B) & P6.0B As of April 30, 2011	SUMMARY (By Region) - CAR, I, ARIIP, II, MARIIS, III, UPRIIS, IV-A, IV-B, V VISAYAS - VI, VII, VIII MINDANAO - IX, X, XI, XII, XIII By region, province and irrigation system/project CAR, I, ARIIP, II, MARIIS, III, III (2), UPRIIS, IV (MLUB), IV (CP), V, VI, VII, VIII, IX, X, XI, XII, XIII

Annex D. Rehabilitation/improvement cost by component (PhP, x10³)

Agos RIS

Year	Project fund source	Allocation	Diversion work	Canal lining	Canal structures	Canal desilting	Canali-zation	Road	Drainage system	Erosion control	Field office	O&M
1995	IOSP II	215	-	-	-	-	-	-	-	182	-	-
1996	IOSP II	3,728	-	3,125	-	-	-	-	-	64	-	-
1998	IOSP II	19,664	-	16,756	-	-	-	-	-	-	-	-
1999	IOSP II GAA	5,273	-	3,481	760	-	-	28	-	-	-	-
2000	Typhoon Damage	200	179	-	-	-	-	-	-	-	-	-
2001	RRNIS 2001 MOOE 2001	365	189	-	-	-	-	-	-	-	-	120
2002	RRENIS 2002	1,000	317	302	-	76	-	-	-	-	179	-
2003	RRENIS 2003, BSSP 2003, RRNIS 2003	2,612	518	1,700			101	-	-	-	-	-
2004	GAA 2003 RA 9206	5,000	-	-	-	-	-	4,502	-	-	-	-
2005	GAA, MOOE, Fund 101, RRENIS	9,543	6,099	-	-	2,413	-	-	-	-	129	-
2006	GAA, NDC 1 2006, NDC 2 2006	11,000	764	8,735	-	136	-	-	-	-	-	-
2007	GAA/RRNIS 2007, Fund 101, CARE	7,000	-	5,268	-	413	-	-	-	-	-	-

Modernisation strategy for the national irrigation systems in the Philippines

Agos RIS. *Continued*

Year	Project fund source	Allocation	Diversion work	Canal lining	Canal structures	Canal desilting	Canali- zation	Road	Drainage system	Erosion control	Field office	O&M
2008	GAA 2008	1,000	-	-	164	680	-	-	-	-	-	-
2009	NDC 5 YR 2009	8,000	-	6,188	-	-	-	-	420	-	-	-
2010	NDC 6 YR 2010, 6.524B YR 2010	19,000	-	15,420	-	402	-	-	-	-	-	-
2011	WB/GAA, CRRENIS, Fund 101	33,045	2,130	10,043	6,501	-	3,043	-	-	-	-	-
2012	Fund 101	3,171	67	484	13	184	2,087	-	-	-	-	-
2013	NISCIS Fund 101, RREIS 101	17,733	-	9,859	49	3,142	-	3,882	325	-	-	-
2014	Fund 101	652	-	-	-	-	-	-	-	-	-	-
2015	RRENIS Fund 101	3,000	-	2,766	78	-	-	-	-	-	-	-

Agos RIS

Year	Project fund source	IMT	IDP	Mapping	FSDE	Equipment	GESA	Mgt fee	Overhead	Contingency	Reserved fund	Local taxes
1995	IOSP II	-	-	-	-	-	-	-	33	-	-	-
1996	IOSP II	-	-	-	-	-	-	-	226	313	-	-
1998	IOSP II	-	-	-	-	-	-	983	1,173	751	-	-
1999	IOSP II GAA	-	-	-	-	-	-	247	270	228	-	-
2000	Typhoon Damage	-	-	-	-	-	-	10	11	-	-	-
2001	RRNIS 2001 MOOE 2001	-	-	-	-	-	-	18	17	-	20	-
2002	RRENIS 2002	-	26	-	-	-	-	50	50	-	-	-
2003	RRENIS 2003, BSSP 2003, RRNIS 2003	-	65				69	90	29	39	-	-
2004	GAA 2003 RA 9206	-	-	-	-	-	-	250	248	-	-	-
2005	GAA, MOOE, Fund 101, RRENIS	-	-	-	-	-	475	427	-	-	-	-
2006	GAA, NDC 1 2006, NDC 2 2006	-	270	-	-	-	-	550	545	-	-	-
2007	GAA/RRNIS 2007, Fund 101, CARE	-	316	-	316	-	-	350	338	-	-	-
2008	GAA 2008	-	28	-	46	-	-	50	32	-	-	-
2009	NDC 5 YR 2009	-	367	-	367	-	-	400	257	-	-	-

Modernisation strategy for the national irrigation systems in the Philippines

Agos RIS. Continued

Year	Project fund source	IMT	IDP	Mapping	FSDE	Equipment	GESA	Mgt fee	Overhead	Contingency	Reserved fund	Local taxes
2010	NDC 6 YR 2010, 6.524B YR 2010	-	823	-	795	-	-	950	610	-	-	-
2011	WB/GAA, CRRENIS, Fund 101	3,675	-	-	363	542	872	1,246	-	3,127	-	1,502
2012	Fund 101	-	88	-	-	-	-	159	90	-	-	-
2013	NISCIS Fund 101, RREIS 101	-	288	-	190	-	-	-	-	-	-	-
2014	Fund 101	-	-	652	-	-	-	-	-	-	-	-
2015	RRENIS Fund 101	-	60	-	7	-	-	-	-	-	-	-

Rehabilitiation/improvement cost by component (PhP, x10^3)

Balanac RIS

Year	Project fund source	Project allocation	Balanac allocation	Diversion work	Canal lining	Canal structure	Canal desilting	Road	Field office	O&M
1995										
1996	IOSP II, Deferred Urgent Repair	758	758	740	-		-	-	-	-
1997	IOSP II, Deferred Urgent Repair, WRDP Y1	6,739	6,739	-	3,874	2,644	81	-	-	-
1998	WRDP YR 2	9,347	9,347	-	8,834	77	-	-	-	-
1999	WRDP YR 3, IOSP Urgent Repair POW 1999	17,389	16,890	6,344	8,515	409	-	271	-	-
2000	IOSP II Urgent Repair, WRDP	23,877	23,114	18,867	-	34	-	-	-	-
2001	GAA, RRNIS, MOOE	1,071	514	212	-	-	62	-	29	117
2002	GAA, RRNEIS El Niño YR 2002	1,730	1,309	-	1,075	-	64	-	-	-
2003	GAA	5,000	1,171	-	-	-	-	1,054	-	-
2004	-	-	-	-	-	-	-	-	-	-
2005	RRENIS, RRNIS	1,432	1,432	401	783	-	94	-	-	-
2006	RRENIS	6,022	6,022	275	654	2,447	-	1,888	-	-
2007	1.5B Augmentation Fund, BSPP	8,050	8,0505	-	3,350	1,500	505	1,418	-	-

Modernisation strategy for the national irrigation systems in the Philippines

Balanac RIS. *Continued*

Year	Project fund Source	Project allocation	Balanac allocation	Diversion work	Canal lining	Canal structure	Canal desilting	Road	Field office	O&M
2008	RRENIS, 1.65B, GESA	6,000	6,0005	-	4,258	-	678	-	94	-
2009	GAA	900	9005	-	534	286	77	-	-	-
2010	NDC 6, GA, 6.524B	49,600	49,6005	14,356	30,506	134	-	-	-	-
2011	BSPP 2011, PIDP	19,255	19,255	-	13,047	113	-	-	-	156
2012	RREIS	10,141	10,141	-	9,065	-	-	-	-	-
2013	NIS/CIS Extn, Fund 101, GAA	72,425	72,425	47,622	18,050	-	-	5,351	-	-
2014	Fund 501 CY 2014, NIS/CIS Extn CY 2014	20,000	20,000	9,789	9,302	2,490	-	-	-	-
2015	Climate change	5,000	4,370	4,144	-	-	-	-	-	-

Rehabilitation/improvement cost by component (PhP, x10³)

Balanac RIS

Year	Project fund source	Training	IMT	Social/envt safeguards	Taxes	IDP	FSDE	Equipment	Mgt fee	Overhead	GESA	Reserved fund
1995												
1996	IOSP II, Deferred Urgent Repair	-	-	-	-	-	-	-	-	7	12	-
1997	IOSP II, Deferred Urgent Repair, WRDP Y1	-	-	-	-	-	-	-	-	16	124	-
1998	WRDP YR 2	-	-	-	-	-	-	-	-	-	435	-
1999	WRDP YR 3, IOSP Urgent Repair POW 1999	-	-	-	-	-	-	-	15	627	710	-
2000	IOSP II Urgent Repair, WRDP	-	-	-	-	-	-	-	1,106	784	1,323	-
2001	GAA, RRNIS, MOOE	-	-	-	-	-	-	-	26	-	17	51
2002	GAA, RRNEIS El Niño YR 2002	-	-	-	-	32	-	-	65	-	64	8
2003	GAA	-	-	-	-	-	-	-	59	58	-	-
2004	-	-	-	-	-	-	-	-	-	-	-	-
2005	RRENIS, RRNIS	-	-	-	-	11	-	-	44	28	71	-
2006	RRENIS	-	-	-	-	159	-	-	301	298	-	-
2007	1.5B Augmentation Fund, BSPP	-	-	-	-	277	125	-	390	268	-	-

Balanac RIS. *Continued*

Year	Project fund source	Training	IMT	Social/ envt safeguards	Taxes	IDP	FSDE	Equipment	Mgt fee	Overhead	GESA	Reserved fund
2008	RRENIS, 1.65B, GESA	-	-	-	-	207	334	-	150	96	-	-
2009	GAA	-	-	-	-	-	-	-	-	-	-	-
2010	NDC 6, GA, 6.524B	-	-	-	-	815	1,359	-	1,480	935	16	-
2011	BSPP 2011, PIDP	387	2,379	13	301	277	536	542	626	523	357	-
2012	RREIS	-	-	-	-	187	-	93	507	289	-	-
2013	NIS/CIS Extn, Fund 101, GAA	-	-	-	-	1,382	20	-	-	-	-	-
2014	Fund 501 CY 2014, NIS/CIS Extn CY 2014					400	20		-	-	-	-
2015	Climate change	-	-	-	-	87	8	-	-	-	-	-

Rehabilitation/improvement cost by component (PhP, x10³)

Sta. Maria RIS

	Fund source	Project allocation	Sta. Maria allocation	Diversion	Canal lining	Canal structure	Canal desilting	Canal repair	Road	Field office	STW	Drainage system
1993	IOSP II	105	105	-	-	-	0	-	102	-	-	-
1994	GAA	445	292	-	-	-	240	-	-	-	-	-
1995	GAA, Local Fund, IOSP II	1,524	1,461	-	750	-	-	-	504	50	-	-
1996	GAA, IOSP II	8,909	7,554	2,088	1,782	286	122	-	1,825	23	-	-
1997	GAA, IOSP	14,909	14,612	-	7,505	47	-	-	3,349	211	1,268	101
1998	IOSP II	16,592	15,187	-	11,795	-	169	-	504	-	-	-
1999	IOSP II, GAA	6,730	5,393	1,250	2,625	-	257	-	-	295	-	-
2000	IOSP II	4,743	3,876	-	3,261	180	-	-	-	-	-	-
2001	GAA	487	487	-	-	-	-	-	-	-	-	-
2002	-	-	-	-	-	-	-	-	-	-	-	-
2003	GAA RRENIS	5,377	4,562	286	155	-	-	-	3,245	-	-	369
2004	GAA 2002, RA 9162	5,000	5,000	-	239	-	-	-	4,263	-	-	-
2005	GAA, RRENIS	2,500	1,245	900	221	-	-	-	-	-	-	-
2006	NDC 1	500	259	-	-	-	-	-	233	-	-	-
2007	RRNIS	1,000	1,000	70	627	5	113	0	0	-	-	-
2008	1.5B Fund, 1.65B 2008, RRNIS, IDP	10,500	10,394	-	5,414	-	315	0	3,134	-	-	24

Modernisation strategy for the national irrigation systems in the Philippines

Sta. Maria RIS. *Continued*

	Fund source	Project allocation	Sta. Maria allocation	Diversion	Canal lining	Canal structure	Canal desilting	Canal repair	Road	Field office	STW	Drainage system
2009	IDP NDC 501, Fund 101, 6B 2009, RRENIS	5,750	5,384	-	4,726	52	-	-	-	-	-	-
2010	NDC 6, 6.524B, 6.524B 2010	9,500	7,942	3,640	2,722	342	-	-	-	-	-	-
2011	BSPP, GAA, PIDP	20,972	19,997	-	11,874	2,982	-	-	-	-	-	829
2012	-	-	-	-	-	-	-	-	-	-	-	-
2013	-	-	-	-	-	-	-	-	-	-	-	-
2014	RREIS CY 2014	10,000	10,000	9,789	-	-	-	-	-	-	-	-
2015	-	-	-	-	-	-	-	-	-	-	-	-

Rehabilitation/improvement cost by component (PhP, x10³)

Sta. Maria RIS

	Fund source	Erosion control	O&M/ MOOE	Training	IMT devt	IDP	FSDE	GESA	Overhead	Contingency	Mgt fee	Reserved fund
1993	IOSP II	-	-	-	-	-	-	-	3	-	-	-
1994	GAA	-	46	-	-	-	-	-	6	-	-	-
1995	GAA, Local Fund, IOSP II	112	-	-	-	-	-	-	46	-	-	-
1996	GAA, IOSP II	-	201	-	-	-	145	268	73	382	360	-
1997	GAA, IOSP	-	-	-	-	-	-		802	700	628	-
1998	IOSP II	-	-	-	-	-	-	-	842	1,152	725	-
1999	IOSP II, GAA	-	199	-	-	-	-	-	285	240	242	-
2000	IOSP II	-	-	-	-	-	-	-	241	-	194	-
2001	GAA	-	414	-	-	-	-	-	-		24	49
2002	-	-	-	-	-	-	-	-	-	-	-	-
2003	GAA RRENIS	-	-	-	-	-	-	-	222	-	228	57
2004	GAA 2002, RA 9162	-	-	-	-	-	-	-	248	-	250	-
2005	GAA, RRENIS	-	-	-	-	-	-	-	62	-	62	-
2006	NDC 1	-	-	-	-	-	-	-	13	-	13	-
2007	RRNIS	-	-	-	-	45	45	-	45	-	50	-
2008	1.5B Fund, 1.65B 2008, RRNIS, IDP	-	-	-	-	307	414	-	338	-	447	-

Modernisation strategy for the national irrigation systems in the Philippines

Sta. Maria RIS. *Continued*

	Fund source	Erosion control	O&M/ MOOE	Training	IMT devt	IDP	FSDE	GESA	Overhead	Contingency	Mgt fee	Reserved fund
2009	IDP NDC 501, Fund 101, 6B 2009, RRENIS	-	-	-	-	114	162	-	151	-	179	-
2010	NDC 6, 6.524B, 6.524B 2010	-	-	-	-	241	345	255	147	-	250	-
2011	BSPP, GAA, PIDP	-	141	279	1,448	279	340	50	663	-	822	-
2012	-	-	-	-	-	-	-	-	-	-	-	-
2013	-	-	-	-	-	-	-	-	-	-	-	-
2014	RREIS CY 2014	-	-	-	-	200	11	-	-	-	-	-
2015	-	-	-	-	-	-	-	-	-	-	-	-

Annex E1. Locations of direct offtakes and drainage inlet along the main canal of Balanac RIS

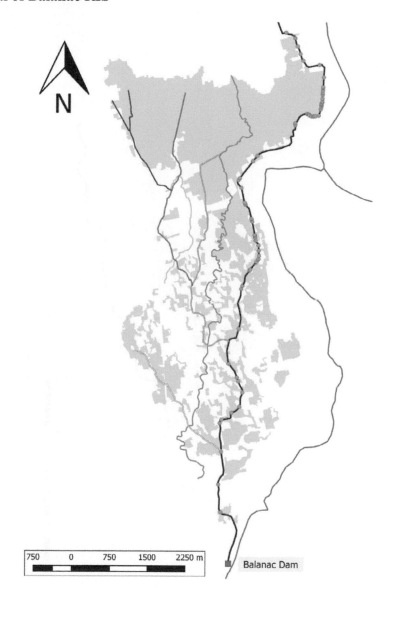

N

| 750 | 0 | 750 | 1500 | 2250 m |

Balanac Dam

Annex E2. Locations of direct offtakes and drainage inlet along the main canal of Sta. Maria RIS

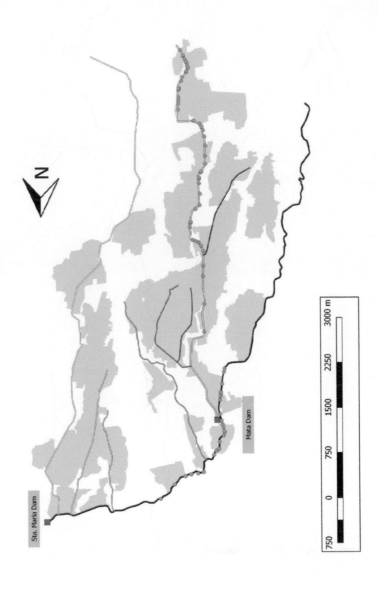

Annex F1. RAP internal indicators for Balanac RIS

Project name:							
	Balanac River Irrigation Systems						
Date:							
01/00/00							

Points for understanding this Indicator Summary

1. This spreadsheet only applies to INTERNAL indicators. A separate spreadsheet is used for EXTERNAL indicators such as Irrigation Efficiency and Relative Water Supply.

2. The majority of the values on this worksheet are automatically transferred from previous worksheets in this spreadsheet.

3. Some of the indicator values on this worksheet must be assigned by the user.

4. The organization of this worksheet is as follows:

a. The alpha-numeric label for each indicator is found in Column A

b. The Primary Indicator name is given in Column B

c. The Sub-Indicator is described in Column C

d. The assigned value for each Sub-Indicator is found in Column D. Also, computed values for each Primary Indicators are found here.

e. The weight assigned to each Sub-Indicator is given in Column E.

f. The original indicator labels, as found in FAO Water Reports 19, are given here.

g. The worksheet in which the original data were entered is given.

Modernisation strategy for the national irrigation systems in the Philippines

Annex F1. *Continued*

Indicator Label	Primary indicator name	Sub-indicator name	Value (0-4)	Weight ing factor	Old indicator label[2]	Worksheet location
		SERVICE and SOCIAL ORDER				
I-1	**Actual** Water Delivery Service to Individual Ownership Units (e.g., field or farm)		1.1		I-1	Final deliveries
I-1A		Measurement of volumes	0.0	1.0	I-1A	
I-1B		Flexibility	0.0	2.0	I-1B	
I-1C		Reliability	1.0	4.0	I-1C	
I-1D		Apparent equity	2.0	4.0	I-1D	
I-2	**Stated** Water Delivery Service to Individual Ownership Units (e.g., field or farm)		2.0		I-5	Project office questions
I-2A		Measurement of volumes	2.0	1.0	I-5A	
I-2B		Flexibility	2.0	2.0	I-5B	
I-2C		Reliability	2.0	4.0	I-5C	
I-2D		Apparent equity	2.0	4.0	I-5D	

---

[2] *(FAO Water Reports)*

Annex F1. *Continued*

Indicator label	Primary indicator name	Sub-indicator name	Value (0-4)	Weight ing factor	Old indicator label	Worksheet location
I-3	**Actual** Water Delivery Service at the most downstream point in the system operated by a paid employee		**0.9**		*I-3*	Final deliveries
I-3A		Number of fields downstream of this point	0.0	1.0	*I-3A*	
I-3B		Measurement of volumes	0.0	4.0	*I-3B*	
I-3C		Flexibility	1.0	4.0	*I-3C*	
I-3D		Reliability	1.0	4.0	*I-3D*	
I-3E		Apparent equity	2.0	4.0	*I-3E*	
I-4	**Stated** Water Delivery Service at the most downstream point in the system operated by a paid employee		**1.9**		*I-7*	Project office questions
I-4A		Number of fields downstream of this point	1.0	1.0	*I-7A*	
I-4B		Measurement of volumes	1.0	4.0	*I-7B*	
I-4C		Flexibility	3.0	4.0	*I-7C*	
I-4D		Reliability	2.0	4.0	*I-7D*	
I-4E		Apparent equity	2.0	4.0	*I-7E*	

Modernisation strategy for the national irrigation systems in the Philippines

Annex F1. *Continued*

Indicator label	Primary indicator name	Sub-indicator name	Value (0-4)	Weighting factor	Old indicator label	Worksheet location
I-5	**Actual** Water Delivery Service by the main canals to the second level canals		**0.4**		*I-4*	Main canal
I-5A		Flexibility	1.0	1.0	*I-4A*	
I-5B		Reliability	0.0	1.0	*I-4B*	
I-5C		Equity	1.0	1.0	*I-4C*	
I-5D		Control of flow rates to the submain as stated	0.0	1.5	*I-4D*	
I-6	**Stated** Water Delivery Service by the main canals to the second level canals		**1.8**		*I-8*	Proj. office Questions
I-6A		Flexibility	1.0	1.0	*I-8A*	
I-6B		Reliability	2.0	1.0	*I-8B*	
I-6C		Equity	2.0	1.0	*I-8C*	
I-6D		Control of flow rates to the submain as stated	2.0	1.5	*I-8D*	
I-7	Social "Order" in the Canal System operated by paid employees		**0.5**		*I-9*	Final deliveries
I-7A		Degree to which deliveries are **NOT** taken when not allowed, or at flow rates greater than allowed	1.0	2.0	*I-9A*	
I-7B		Noticeable **non**-existence of unauthorized turnouts from canals.	0.0	1.0	*I-9B*	
I-7C		Lack of vandalism of structures	0.0	1.0	*I-9C*	

Annex F1. *Continued*

Indicator label	Primary indicator name	Sub-indicator name	Value (0-4)	Weighting factor	Old indicator label	Worksheet location
	MAIN CANAL					
I-8		Cross regulator hardware (main canal)	1.6		I-10	Main canal
I-8A		Ease of cross regulator operation under the current target operation. This does not mean that the current targets are being met; rather this rating indicates how easy or difficult it would be to move the cross regulators to meet the targets	0.0	1.0	I-10A	
I-8B		Level of maintenance of the cross regulators	0.0	1.0	I-10C	
I-8C		Lack of water level fluctuation	1.0	3.0	I-10D	
I-8D		Travel time of a flow rate change throughout this canal level	4.0	2.0	I-10E	
I-9		Turnouts from the main canal	0.7		I-12	Main canal
I-9A		Ease of turnout operation under the current target operation. This does not mean that the current targets are being met; rather this rating indicates how easy or difficult it would be to move the turnouts and measure flows to meet the targets	0.0	1.0	I-12A	
I-9B		Level of maintenance	0.0	1.0	I-12C	
I-9C		Flow rate capacities	2.0	1.0	I-12D	

Modernisation strategy for the national irrigation systems in the Philippines

Annex F1. *Continued*

Indicator label	Primary indicator name	Sub-indicator name	Value (0-4)	Weighting factor	Old indicator label	Worksheet location
I-10		Regulating Reservoirs in the main canal	**0.0**		*I-13*	Main canal
I-10A		Suitability of the number of location(s)	0.0	2.0	*I-13A*	
I-10B		Effectiveness of operation	0.0	2.0	*I-13B*	
I-10C		Suitability of the storage/buffer capacities	0.0	1.0	*I-13C*	
I-10D		Maintenance	0.0	1.0	*I-13D*	
I-11		Communications for the main canal	**1.0**		*I-14*	Main canal
I-11A		Frequency of communications with the next higher level? (hr)	1.0	2.0	*I-14A*	
I-11B		Frequency of communications by operators or supervisors with their customers	1.0	2.0	*I-14B*	
I-11C		Dependability of voice communications by phone or radio.	1.0	3.0	*I-14C*	
I-11D		Frequency of visits by upper level supervisors to the field.	0.0	1.0	*I-14D*	
I-11E		Existence and frequency of remote monitoring (either automatic or manual) at key **spill** points, including the end of the canal	0.0	1.0	*I-14E*	
I-11F		Availability of roads along the canal	2.0	2.0	*I-14F*	

Annex F1. *Continued*

Indicator label	Primary indicator name	Sub-indicator name	Value (0-4)	Weight ing factor	*Old indicator label*	*Worksheet location*
I-12		General Conditions for the main canal	**1.4**		*I-15*	Main canal
I-12A		General level of maintenance of the canal floor and canal banks	2.0	1.0	*I-15A*	
I-12B		General lack of underused seepage (note: if deliberate conjunctive use is practiced, some seepage may be desired)	1.0	1.0	*I-15B*	
I-12C		Availability of proper equipment and staff to adequately maintain this canal	0.0	2.0	*I-15C*	
I-12D		Travel time from the maintenance yard to the most distant point along this canal (for crews and maintenance equipment)	4.0	1.0	*I-15D*	
I-13		Operation of the main canal	**1.3**		*I-16*	Main canal
I-13A		How frequently does the headworks respond to realistic real time feedback from the operators/observers of this canal level? This question deals with a mismatch of orders, and problems associated with wedge storage variations and wave travel times	2.7	2.0	*I-16A*	
I-13B		Existence and effectiveness of water ordering/delivery procedures to match actual demands. This is different than the previous question, because the previous question dealt with problems that occur AFTER a change has been made	0.0	1.0	*I-16B*	

Annex F1. *Continued*

Indicator label	Primary indicator name	Sub-indicator name	Value (0-4)	Weighting factor	Old indicator label	Worksheet location
I-13C		Clarity and correctness of instructions to operators	1.3	1.0	*I-16C*	
I-13D		How frequently is the whole length of this canal checked for problems and reported to the office? This means one or more persons physically drive all the sections of the canal	0.0	1.0	*I-16D*	
	Second level canals					
I-14		Cross regulator hardware (second level canals)	1.1		*I-10*	Second level canals
I-14A		Ease of cross regulator operation under the current target operation. This does not mean that the current targets are being met; rather this rating indicates how easy or difficult it would be to move the cross regulators to meet the targets	0.0	1.0	*I-10A*	
I-14B		Level of maintenance of the cross regulators	0.0	1.0	*I-10C*	

Annex F1. *Continued*

Indicator label	Primary indicator name	Sub-indicator name	Value (0-4)	Weighting factor	Old indicator label	Worksheet location
I-14C		Lack of water level fluctuation	0.0	3.0	I-10D	
I-14D		Travel time of a flow rate change throughout this canal level	4.0	2.0	I-10E	
I-15		Turnouts from the second level canals	**1.3**		I-12	Second level canals
I-15A		Ease of turnout operation under the current target operation. This does not mean that the current targets are being met; rather this rating indicates how easy or difficult it would be to move the turnouts and measure flows to meet the targets	0.0	1.0	I-12A	
I-15B		Level of maintenance	0.0	1.0	I-12C	
I-15C		Flow rate capacities	4.0	1.0	I-12D	
I-16		Regulating Reservoirs in the second level canals	**0.0**		I-13	Second level canals
I-16A		Suitability of the number of location(s)	0.0	2.0	I-13A	
I-16B		Effectiveness of operation	0.0	2.0	I-13B	

Modernisation strategy for the national irrigation systems in the Philippines

Annex F1. *Continued*

Indicator label	Primary indicator name	Sub-indicator name	Value (0-4)	Weighting factor	Old indicator label	Worksheet location
I-16C		Suitability of the storage/buffer capacities	0.0	1.0	I-13C	
I-16D		Maintenance	0.0	1.0	I-13D	
I-17		Communications for the second level canals	1.3		I-I20	Second level canals
I-17A		Frequency of communications with the next higher level? (hr)	1.0	2.0	I-20A	
I-17B		Frequency of communications by operators or supervisors with their customers	1.0	2.0	I-20B	
I-17C		Dependability of voice communications by phone or radio	2.0	3.0	I-20C	
I-17D		Frequency of visits by upper level supervisors to the field	0.0	1.0	I-20D	
I-17E		Existence and frequency of remote monitoring (either automatic or manual) at key **spill** points, including the end of the canal	0.0	1.0	I-20E	
I-17F		Availability of roads along the canal	2.0	2.0	I-21F	

Annex F1. *Continued*

Indicator label	Primary indicator name	Sub-indicator name	Value (0-4)	Weight ing factor	Old indicator label	Worksheet location
I-18		General Conditions for the second level canals	1.2		I-21	Second level canals
I-18A		General level of maintenance of the canal floor and canal banks	1.0	1.0	I-21B	
I-18B		General lack of <u>undesired</u> seepage (note: if deliberate conjunctive use is practiced, some seepage may be desired)	1.0	1.0	I-21C	
I-18C		Availability of proper equipment and staff to adequately maintain this canal	0.0	2.0	I-21D	
I-18D		Travel time from the maintenance yard to the most distant point along this canal (for crews and maintenance equipment)	4.0	1.0	I-21E	

Modernisation strategy for the national irrigation systems in the Philippines

Annex F1. *Continued*

Indicator label	Primary indicator name	Sub-indicator name	Value (0-4)	Weight ing factor	Old indicator label	Worksheet location
I-19		Operation of the second level canals	1.3		I-22	Second level canals
I-19A		How frequently does the headworks respond to realistic real time feedback from the operators/observers of this canal level? This question deals with a mismatch of orders, and problems associated with wedge storage variations and wave travel times	2.7	2.0	I-22A	
I-19B		Existence and effectiveness of water ordering/delivery procedures to match actual demands. This is different than the previous question, because the previous question dealt with problems that occur AFTER a change has been made	0.0	1.0	I-22B	
I-19C		Clarity and correctness of instructions to operators	1.3	1.0	I-22C	
I-19D		How frequently is the whole length of this canal checked for problems and reported to the office? This means one or more persons physically drive all the sections of the canal	0.0	1.0	I-22D	

Annex F1. *Continued*

Indicator label	Primary indicator name	Sub-indicator name	Value (0-4)	Weighting factor	Old indicator label	Worksheet location
	Third level canals					
I-20		Cross regulator hardware (third level canals)	1.1			Third level canals
I-20A		Ease of cross regulator operation under the current target operation. This does not mean that the current targets are being met; rather this rating indicates how easy or difficult it would be to move the cross regulators to meet the targets	0.0	1.0		
I-20B		Level of maintenance of the cross regulators	0.0	1.0		
I-20C		Lack of water level fluctuation	0.0	3.0		
I-20D		Travel time of a flow rate change throughout this canal level	4.0	2.0		

Modernisation strategy for the national irrigation systems in the Philippines

Annex F1. *Continued*

Indicator label	Primary indicator name	Sub-indicator name	Value (0-4)	Weighting factor	Old indicator label	Worksheet location
I-21		Turnouts from the third level canals	**1.3**			Third level canals
I-21A		Ease of turnout operation under the current target operation. This does not mean that the current targets are being met; rather this rating indicates how easy or difficult it would be to move the turnouts and measure flows to meet the targets	0.0	1.0		
I-21B		Level of maintenance	0.0	1.0		
I-21C		Flow rate capacities	4.0	1.0		
I-22		Regulating Reservoirs in the third level canals	**0.0**			Third level canals
I-22A		Suitability of the number of location(s)	0.0	2.0		
I-22B		Effectiveness of operation	0.0	2.0		
I-22C		Suitability of the storage/buffer capacities	0.0	1.0		
I-22D		Maintenance	0.0	1.0		

Annex F1. *Continued*

Indicator label	Primary indicator name	Sub-indicator name	Value (0-4)	Weighting factor	Old indicator label	Worksheet location
I-23		Communications for the third level canals	1.3			Third level canals
I-23A		Frequency of communications with the next higher level? (hr)	1.0	2.0		
I-23B		Frequency of communications by operators or supervisors with their customers	1.0	2.0		
I-23C		Dependability of voice communications by phone or radio	2.0	3.0		
I-23D		Frequency of visits by upper level supervisors to the field	0.0	1.0		
I-23E		Existence and frequency of remote monitoring (either automatic or manual) at key **spill** points, including the end of the canal	0.0	1.0		
I-23F		Availability of roads along the canal	2.0	2.0		
I-24		General Conditions for the third level canals	1.2			Third level canals
I-24A		General level of maintenance of the canal floor and canal banks	1.0	1.0		
I-24B		General lack of underdesired seepage (note: if deliberate conjunctive use is practiced, some seepage may be desired)	1.0	1.0		

Modernisation strategy for the national irrigation systems in the Philippines

Annex F1. *Continued*

Indicator label	Primary indicator name	Sub-indicator name	Value (0-4)	Weight ing factor	Old indicator label	Worksheet location
I-24C		Availability of proper equipment and staff to adequately maintain this canal	0.0	2.0		
I-24D		Travel time from the maintenance yard to the most distant point along this canal (for crews and maintenance equipment)	4.0	1.0		
I-25		Operation of the third level canals	**0.8**			Third level canals
I-25A		How frequently does the headworks respond to realistic real time feedback from the operators/observers of this canal level? This question deals with a mismatch of orders, and problems associated with wedge storage variations and wave travel times	1.3	2.0		
I-25B		Existence and effectiveness of water ordering/delivery procedures to match actual demands. This is different than the previous question, because the previous question dealt with problems that occur AFTER a change has been made	0.0	1.0		
I-25C		Clarity and correctness of instructions to operators	1.3	1.0		

Annex F1. *Continued*

Indicator label	Primary indicator name	Sub-indicator name	Value (0-4)	Weighting factor	Old indicator label	Worksheet location
I-25D		How frequently is the whole length of this canal checked for problems and reported to the office? This means one or more persons physically drive all the sections of the canal	0.0	1.0		
		Budgets, Employees, WUAs				
I-26	Budgets		1.2		I-23	Project office questions
I-26A		What percentage of the total project (including WUA) Operation and Maintenance (O&M) is collected as in-kind services, and/or water fees from water users?	1.0	2.0	I-23A	
I-26B		Adequacy of the actual dollars and in-kind services that is available (from all sources) to sustain adequate Operation and Maintenance (O&M) with the present mode of operation	0.0	2.0	I-23B	
I-26C		Adequacy of spending on modernization of water delivery structures (as contrasted to rehabilitation or regular operation)	4.0	1.0	I-23C	

Modernisation strategy for the national irrigation systems in the Philippines

Annex F1. *Continued*

Indicator label	Primary indicator name	Sub-indicator name	Value (0-4)	Weighting factor	Old indicator label	Worksheet location
I-27	Employees		**1.9**		*I-24*	Project Employees
I-27A		Frequency and adequacy of training of operators and middle managers (not secretaries and drivers). This should include employees at all levels of the distribution system, not only those who work in the office	2.0	1.0	*I-24A*	
I-27B		Availability of written performance rules	1.0	1.0	*I-24B*	
I-27C		Power of employees to make decisions	3.0	2.5	*I-24C*	
I-27D		Ability of the project to dismiss employees with cause.	2.0	2.0	*I-24D*	
I-27E		Rewards for enemplary service	2.0	1.0	*I-24E*	
I-27F		Relative salary of an operator compared to a day laborer	1.0	2.0	*I-24F*	
I-28		Water User Associations	**0.3**		*I-25*	WUA
I-28A		Percentage of all project users who have a functional, formal unit that participates in water distribution	0.0	2.5	*I-25A*	

Annex F1. *Continued*

Indicator label	Primary indicator name	Sub-indicator name	Value (0-4)	Weight ing factor	Old indicator label	Worksheet location
I-28B		Actual ability of the strong Water User Associations to influence real-time water deliveries to the WUA	0.0	1.0	I-25B	
I-28C		Ability of the WUA to rely on effective outside help for enforcement of its rules	0.0	1.0	I-25C	
I-28D		Legal basis for the WUAs	1.0	1.0	I-25D	
I-28E		Financial strength of WUAS	1.0	1.0	I-25E	
I-29	Mobility and Size of Operations Staff	Operation staff mobility and efficiency, based on the ratio of operating staff to the number of turnouts	0.0		I-28	Project office questions
I-30	Computers for billing and record management	The extent to which computers are used for billing and record management	0.0		I-30	Project office questions
I-31	Computers for canal control	The extent to which computers (either central or on-site) are used for canal control	0.0		I-31	Project office questions

Annex F1. *Continued*

Indicator label	Primary indicator name	Sub-indicator name	Value (0–4)	Weight ing factor	Old indicator label	Worksheet location
	INDICATORS THAT WERE NOT PREVIOUSLY COMPUTED	**THESE INDICATORS REQUIRE THE INPUT OF VALUES (0–4) IN EACH OF THE BOXES**				
I-32	Ability of the present water delivery service to individual fields, to support pressurized irrigation methods		**0.0**		*I-26*	n/a
I-32A	Measurement and control of volumes to the field	4 - Excellent volumetric metering and control; 3.5 - Ability to measure flow rates reasonably well, but not volume. Flow is well controlled; 2.5 - Cannot measure flow, but can control flow rates; 0 - Cannot control the flow rate, even though it can be measured	**0.0**	1.0	*I-26A*	n/a
I-32B	Flexibility to the field	4 - Arranged delivery, with frequency, rate and duration promised. All can be varied upon request; 3 - Same as 4, but cannot vary the duration; 2 - 2 variables are fixed, but arranged schedule; 0 - Rotation	**0.0**	1.0	*I-26B*	n/a

Annex F1. *Continued*

Indicator label	Primary indicator name	Sub-indicator name	Value (0-4)	Weighting factor	Old indicator label	Worksheet location
I-32C	Reliability to the field	4 - Water always arrives as promised, including the appropriate volume; 3 - A few days of delay occasionally occur, but water is still very reliable in rate and duration; 0 - More than a few days delay.	0.0	1.0	I-26C	n/a
I-33	Changes required to be able to support pressurized irrigation methods		0.0		I-27	n/a
I-33A	Procedures, Management	4 - No changes in water ordering, staff training, or mobility; 3.5 - Improved training, only. The basic procedures/conditions are just fine, they just are not being implemented to their full extent; 3.0 - Minor changes in water ordering, mobility, training, incentive programs; 2.0 - Major changes in 1 of the above; 1 - Major changes in 2 of the above; 0 - Need to completely revamp or convert almost everything	0.0	1.0	I-27A	Management

Modernisation strategy for the national irrigation systems in the Philippines

Annex F1. *Continued*

Indicator label	Primary indicator name	Sub-indicator name	Value (0-4)	Weighting factor	Old indicator label	Worksheet location
I-33B	Hardware	4 - No changes needed; 3.5 - Only need to repair some of the existing structures so that they are workable again.; 3.0 - Improved communications, repair of some existing structures, and a few key new structures (less than US$300/ha needed), OR…very little change to existing, but new structures are needed for water recirculation; 2 - Larger capital expenditures - US$ 300 - US$ 600/ha; 1 - Larger capital expenditures needed (up to US$ 1500/ha); 0 - Almost complete reworking of the system is needed	**0.0**	1.0	*I-27B*	Hardware
I-34	Sophistication in receiving and using feedback information. This does not need to be automatic.	4 - Continuous feedback and continuous use of information to change inflows, with all key points monitored. Or, minimal feed back is necessary, such as with closed pipe systems.; 3 - Feedback several times a day and rapid use (within a few hours) of that information, at major points.; 2 - Feedback once/day from key points and appropriate use of information within a day; 1 - Weekly feedback and appropriate usage, or once/day feeback but poor usage of the information; 0 - No meaningful feedback, or else there is a lot of feedback but no usage	**0.0**		*I-29*	n/a

Annex F1. *Continued*

Indicator label	Primary indicator name	Sub-indicator name	Value (0-4)	Weight ing factor	Old indicator label	Worksheet location
		SPECIAL INDICATORS THAT DO NOT HAVE A 0-4 RATING SCALE				
I-35	Turnout density	Number of water users downstream of employee-operated turnouts	109			*Final deliveries*
I-36	Turnouts/Operator	(Number of turnouts operated by paid employees)/(Paid Employees)	4.6			*Project Office*
I-37	Main canal chaos	(Actual/Stated) Overall Service by the main canal	0.25			
I-38	Second level chaos	(Actual/Stated) Overall Service at the most downstream point operated by a paid employee	0.48			
I-39	Field level chaos	(Actual/Stated) Overall Service to the Individual Ownership Units	0.55			

Annex F2. RAP internal indicators for Sta. Maria RIS

Project name:		
	Sta Maria River Irrigation Systems	
Date:		
	01/00/00	

Points for understanding this Indicator Summary

1. This spreadsheet only applies to INTERNAL indicators. A separate spreadsheet is used for EXTERNAL indicators such as Irrigation Efficiency and Relative Water Supply.

2. The majority of the values on this worksheet are automatically transferred from previous worksheets in this spreadsheet.

3. Some of the indicator values on this worksheet must be assigned by the user.

4. The organization of this worksheet is as follows:

a. The alpha-numeric label for each indicator is found in Column A

b. The Primary Indicator name is given in Column B

c. The Sub-Indicator is described in Column C

d. The assigned value for each Sub-Indicator is found in Column D. Also, computed values for each Primary Indicators are found here

e. The weight assigned to each Sub-Indicator is given in Column E

f. The original indicator labels, as found in FAO Water Reports 19, are given here

g. The worksheet in which the original data were entered is given

Modernisation strategy for the national irrigation systems in the Philippines

Annex F2. *Continued*

Indicator label	Primary indicator name	Sub-indicator name	Value (0-4)	Weight ing factor	Old indicator label[3]	Worksheet location
		SERVICE and SOCIAL ORDER				
I-1	**Actual** Water Delivery Service to Individual Ownership Units (e.g., field or farm)		1.1		*I-1*	Final deliveries
I-1A		Measurement of volumes	0.0	1.0	*I-1A*	
I-1B		Flexibility	0.0	2.0	*I-1B*	
I-1C		Reliability	1.0	4.0	*I-1C*	
I-1D		Apparent equity	2.0	4.0	*I-1D*	
I-2	**Stated** Water Delivery Service to Individual Ownership Units (e.g., field or farm)		2.6		*I-5*	Project office questions
I-2A		Measurement of volumes	3.0	1.0	*I-5A*	
I-2B		Flexibility	3.0	2.0	*I-5B*	
I-2C		Reliability	2.0	4.0	*I-5C*	
I-2D		Apparent equity	3.0	4.0	*I-5D*	

---

[3] *(FAO Water Reports)*

Annex F2. *Continued*

Indicator label	Primary indicator name	Sub-indicator name	Value (0-4)	Weighting factor	Old indicator label	Worksheet location
I-3		**Actual** Water Delivery Service at the most downstream point in the system operated by a paid employee	**0.9**		*I-3*	Final deliveries
I-3A		Number of fields downstream of this point	0.0	1.0	*I-3A*	
I-3B		Measurement of volumes	0.0	4.0	*I-3B*	
I-3C		Flexibility	1.0	4.0	*I-3C*	
I-3D		Reliability	1.0	4.0	*I-3D*	
I-3E		Apparent equity	2.0	4.0	*I-3E*	
I-4		**Stated** Water Delivery Service at the most downstream point in the system operated by a paid employee	**2.6**		*I-7*	Project office questions
I-4A		Number of fields downstream of this point	0.0	1.0	*I-7A*	
I-4B		Measurement of volumes	2.0	4.0	*I-7B*	
I-4C		Flexibility	3.0	4.0	*I-7C*	
I-4D		Reliability	4.0	4.0	*I-7D*	
I-4E		Apparent equity	2.0	4.0	*I-7E*	

Modernisation strategy for the national irrigation systems in the Philippines

Annex F2. *Continued*

Indicator label	Primary indicator name	Sub-indicator name	Value (0-4)	Weighting factor	Old indicator label	Worksheet location
I-5		**Actual** Water Delivery Service by the main canals to the second level canals	**1.2**		*I-4*	Main canal
I-5A		Flexibility	1.0	1.0	*I-4A*	
I-5B		Reliability	1.0	1.0	*I-4B*	
I-5C		Equity	2.0	1.0	*I-4C*	
I-5D		Control of flow rates to the submain as stated	1.0	1.5	*I-4D*	
I-6		**Stated** Water Delivery Service by the main canals to the second level canals	**3.0**		*I-8*	Project office questions
I-6A		Flexibility	3.0	1.0	*I-8A*	
I-6B		Reliability	3.0	1.0	*I-8B*	
I-6C		Equity	3.0	1.0	*I-8C*	
I-6D		Control of flow rates to the submain as stated	3.0	1.5	*I-8D*	
I-7		Social "Order" in the Canal System operated by paid employees	**1.5**		*I-9*	Final deliveries
I-7A		Degree to which deliveries are **NOT** taken when not allowed, or at flow rates greater than allowed	2.0	2.0	*I-9A*	

Annex F2. *Continued*

Indicator label	Primary indicator name	Sub-indicator name	Value (0-4)	Weighting factor	Old indicator label	Worksheet location
I-7B		Noticeable **non**-existence of unauthorized turnouts from canals	0.0	1.0	I-9B	
I-7C		Lack of vandalism of structures	2.0	1.0	I-9C	
I-8	**MAIN CANAL**	Cross regulator hardware (main canal)	**3.1**		I-10	Main canal
I-8A		Ease of cross regulator operation under the current target operation. This does not mean that the current targets are being met; rather this rating indicates how easy or difficult it would be to move the cross regulators to meet the targets	3.0	1.0	I-10A	
I-8B		Level of maintenance of the cross regulators	2.0	1.0	I-10C	
I-8C		Lack of water level fluctuation	3.0	3.0	I-10D	
I-8D		Travel time of a flow rate change throughout this canal level	4.0	2.0	I-10E	
I-9		Turnouts from the main canal	**2.3**	1.0	I-12	Main canal
I-9A		Ease of turnout operation under the current target operation. This does not mean that the current targets are being met; rather this rating indicates how easy or difficult it would be to move the turnouts and measure flows to meet the targets	2.0	1.0	I-12A	

Modernisation strategy for the national irrigation systems in the Philippines

Annex F2. *Continued*

Indicator label	Primary indicator name	Sub-indicator name	Value (0-4)	Weighting factor	Old indicator label	Worksheet location
I-9B		Level of maintenance	1.0	1.0	I-12C	
I-9C		Flow rate capacities	4.0	1.0	I-12D	
I-10		Regulating Reservoirs in the main canal	**0.0**		I-13	Main canal
I-10A		Suitability of the number of location(s)	0.0	2.0	I-13A	
I-10B		Effectiveness of operation	0.0	2.0	I-13B	
I-10C		Suitability of the storage/buffer capacities	0.0	1.0	I-13C	
I-10D		Maintenance	0.0	1.0	I-13D	
I-11		Communications for the main canal	**1.5**		I-14	Main canal
I-11A		Frequency of communications with the next higher level? (hr)	1.0	2.0	I-14A	
I-11B		Frequency of communications by operators or supervisors with their customers	1.0	2.0	I-14B	
I-11C		Dependability of voice communications by phone or radio	3.0	3.0	I-14C	
I-11D		Frequency of visits by upper level supervisors to the field	0.0	1.0	I-14D	
I-11E		Existence and frequency of remote monitoring (either automatic or manual) at key **spill** points, including the end of the canal	0.0	1.0	I-14E	
I-11F		Availability of roads along the canal	2.0	2.0	I-14F	

Annex F2. *Continued*

Indicator label	Primary indicator name	Sub-indicator name	Value (0-4)	Weight ing factor	Old indicator label	Worksheet location
I-12		General Conditions for the main canal	1.4		*I-15*	Main canal
I-12A		General level of maintenance of the canal floor and canal banks	2.0	1.0	*I-15A*	
I-12B		General lack of underdesired seepage (note: if deliberate conjunctive use is practiced, some seepage may be desired)	1.0	1.0	*I-15B*	
I-12C		Availability of proper equipment and staff to adequately maintain this canal	0.0	2.0	*I-15C*	
I-12D		Travel time from the maintenance yard to the most distant point along this canal (for crews and maintenance equipment)	4.0	1.0	*I-15D*	
I-13		Operation of the main canal	1.9		*I-16*	Main canal
I-13A		How frequently does the headworks respond to realistic real time feedback from the operators/observers of this canal level? This question deals with a mismatch of orders, and problems associated with wedge storage variations and wave travel times	4.0	2.0	*I-16A*	

Modernisation strategy for the national irrigation systems in the Philippines

Annex F2. *Continued*

Indicator label	Primary indicator name	Sub-indicator name	Value (0-4)	Weight ing factor	Old indicator label	Worksheet location
I-13B		Existence and effectiveness of water ordering/delivery procedures to match actual demands. This is different than the previous question, because the previous question dealt with problems that occur AFTER a change has been made	0.0	1.0	*I-16B*	
I-13C		Clarity and correctness of instructions to operators	1.3	1.0	*I-16C*	
I-13D		How frequently is the whole length of this canal checked for problems and reported to the office? This means one or more persons physically drive all the sections of the canal	0.0	1.0	*I-16D*	
	Second level canals					
I-14		Cross regulator hardware (second level canals)	1.9		*I-10*	*Second level canals*
I-14A		Ease of cross regulator operation under the current target operation. This does not mean that the current targets are being met; rather this rating indicates how easy or difficult it would be to move the cross regulators to meet the targets	3.0	1.0	*I-10A*	

Annex F2. *Continued*

Indicator label	Primary indicator name	Sub-indicator name	Value (0-4)	Weight ing factor	Old indicator label	Worksheet location
I-14B		Level of maintenance of the cross regulators	2.0	1.0	*I-10C*	
I-14C		Lack of water level fluctuation	0.0	3.0	*I-10D*	
I-14D		Travel time of a flow rate change throughout this canal level	4.0	2.0	*I-10E*	
I-15		Turnouts from the second level canals	**1.7**		*I-12*	Second level canals
I-15A		Ease of turnout operation under the current target operation. This does not mean that the current targets are being met; rather this rating indicates how easy or difficult it would be to move the turnouts and measure flows to meet the targets	0.0	1.0	*I-12A*	
I-15B		Level of maintenance	1.0	1.0	*I-12C*	
I-15C		Flow rate capacities	4.0	1.0	*I-12D*	
I-16		Regulating Reservoirs in the second level canals	**0.0**		*I-13*	Second level canals
I-16A		Suitability of the number of location(s)	0.0	2.0	*I-13A*	
I-16B		Effectiveness of operation	0.0	2.0	*I-13B*	
I-16C		Suitability of the storage/buffer capacities	0.0	1.0	*I-13C*	

Modernisation strategy for the national irrigation systems in the Philippines

Annex F2. *Continued*

Indicator label	Primary indicator name	Sub-indicator name	Value (0-4)	Weight ing factor	Old indicator label	Worksheet location
I-16D		Maintenance	0.0	1.0	*I-13D*	
I-17		Communications for the second level canals	**1.3**		*I-120*	Second level canals
I-17A		Frequency of communications with the next higher level? (hr)	1.0	2.0	*I-20A*	
I-17B		Frequency of communications by operators or supervisors with their customers	1.0	2.0	*I-20B*	
I-17C		Dependability of voice communications by phone or radio	2.0	3.0	*I-20C*	
I-17D		Frequency of visits by upper level supervisors to the field	0.0	1.0	*I-20D*	
I-17E		Existence and frequency of remote monitoring (either automatic or manual) at key **spill** points, including the end of the canal	0.0	1.0	*I-20E*	
I-17F		Availability of roads along the canal	2.0	2.0	*I-21F*	
I-18		General Conditions for the second level canals	**1.8**		*I-21*	Second level canals
I-18A		General level of maintenance of the canal floor and canal banks	3.0	1.0	*I-21B*	
I-18B		General lack of undesired seepage (note: if deliberate conjunctive use is practiced, some seepage may be desired)	2.0	1.0	*I-21C*	

Annex F2. *Continued*

Indicator label	Primary indicator name	Sub-indicator name	Value (0-4)	Weight ing factor	Old indicator label	Worksheet location
I-18C		Availability of proper equipment and staff to adequately maintain this canal	0.0	2.0	*I-21D*	
I-18D		Travel time from the maintenance yard to the most distant point along this canal (for crews and maintenance equipment)	4.0	1.0	*I-21E*	
I-19		Operation of the second level canals	**1.9**		*I-22*	Second level canals
I-19A		How frequently does the headworks respond to realistic real time feedback from the operators/observers of this canal level? This question deals with a mismatch of orders, and problems associated with wedge storage variations and wave travel times	4.0	2.0	*I-22A*	
I-19B		Existence and effectiveness of water ordering/delivery procedures to match actual demands. This is different than the previous question, because the previous question dealt with problems that occur AFTER a change has been made	0.0	1.0	*I-22B*	
I-19C		Clarity and correctness of instructions to operators	1.3	1.0	*I-22C*	

Modernisation strategy for the national irrigation systems in the Philippines

Annex F2. *Continued*

Indicator label	Primary indicator name	Sub-indicator name	Value (0-4)	Weighting factor	Old indicator label	Worksheet location
I-19D		How frequently is the whole length of this canal checked for problems and reported to the office? This means one or more persons physically drive all the sections of the canal	0.0	1.0	I-22D	
	Third level canals					
I-20		Cross regulator hardware (third level canals)	#DIV/0!			Third level canals
I-20A		Ease of cross regulator operation under the current target operation. This does not mean that the current targets are being met; rather this rating indicates how easy or difficult it would be to move the cross regulators to meet the targets	0.0	1.0		
I-20B		Level of maintenance of the cross regulators	0.0	1.0		
I-20C		Lack of water level fluctuation	#DIV/0!	3.0		
I-20D		Travel time of a flow rate change throughout this canal level	FALSE	2.0		

Annex F2. *Continued*

Indicator label	Primary indicator name	Sub-indicator name	Value (0-4)	Weight ing factor	Old indicator label	Worksheet location
I-21		Turnouts from the third level canals	**0.0**			Third level canals
I-21A		Ease of turnout operation under the current target operation. This does not mean that the current targets are being met; rather this rating indicates how easy or difficult it would be to move the turnouts and measure flows to meet the targets	0.0	1.0		
I-21B		Level of maintenance	0.0	1.0		
I-21C		Flow rate capacities	0.0	1.0		
I-22		Regulating Reservoirs in the third level canals	**0.0**			Third level canals
I-22A		Suitability of the number of location(s)	0.0	2.0		
I-22B		Effectiveness of operation	0.0	2.0		
I-22C		Suitability of the storage/buffer capacities	0.0	1.0		
I-22D		Maintenance	0.0	1.0		

Modernisation strategy for the national irrigation systems in the Philippines

Annex F2. *Continued*

Indicator label	Primary indicator name	Sub-indicator name	Value (0-4)	Weighting factor	*Old indicator label*	*Worksheet location*
I-23		Communications for the third level canals	**0.0**			Third level canals
I-23A		Frequency of communications with the next <u>higher</u> level? (hr)	FALSE	2.0		
I-23B		Frequency of communications by operators or supervisors with their customers	FALSE	2.0		
I-23C		Dependability of voice communications by phone or radio	0.0	3.0		
I-23D		Frequency of visits by upper level supervisors to the field	FALSE	1.0		
I-23E		Existence and frequency of remote monitoring (either automatic or manual) at key **spill** points, including the end of the canal	0.0	1.0		
I-23F		Availability of roads along the canal	0.0	2.0		
I-24		General Conditions for the third level canals	**0.0**			Third level canals
I-24A		General level of maintenance of the canal floor and canal banks	0.0	1.0		
I-24B		General lack of <u>undesired</u> seepage (note: if deliberate conjunctive use is practiced, some seepage may be desired)	0.0	1.0		

Annex F2. *Continued*

Indicator label	Primary indicator name	Sub-indicator name	Value (0-4)	Weight ing factor	Old indicator label	Worksheet location
I-24C		Availability of proper equipment and staff to adequately maintain this canal	0.0	2.0		
I-24D		Travel time from the maintenance yard to the most distant point along this canal (for crews and maintenance equipment)	FALSE	1.0		
I-25		Operation of the third level canals	**0.0**			Third level canals
I-25A		How frequently does the headworks respond to realistic real time feedback from the operators/observers of this canal level? This question deals with a mismatch of orders, and problems associated with wedge storage variations and wave travel times	0.0	2.0		
I-25B		Existence and effectiveness of water ordering/delivery procedures to match actual demands. This is different than the previous question, because the previous question dealt with problems that occur AFTER a change has been made	0.0	1.0		
I-25C		Clarity and correctness of instructions to operators	0.0	1.0		

Modernisation strategy for the national irrigation systems in the Philippines

Annex F2. *Continued*

Indicator label	Primary indicator name	Sub-indicator name	Value (0-4)	Weighting factor	Old indicator label	Worksheet location
I-25D		How frequently is the whole length of this canal checked for problems and reported to the office? This means one or more persons physically drive all the sections of the canal	0.0	1.0		
		Budgets, Employees, WUAs				
I-26	Budgets		**2.4**		I-23	Project office questions
I-26A		What percentage of the total project (including WUA) Operation and Maintenance (O&M) is collected as in-kind services, and/or water fees from water users?	4.0	2.0	I-23A	
I-26B		Adequacy of the actual dollars and in-kind services that is available (from all sources) to sustain adequate Operation and Maintenance (O&M) with the present mode of operation	0.0	2.0	I-23B	
I-26C		Adequacy of spending on modernization of the water delivery operation/structures (as contrasted to rehabilitation or regular operation)	4.0	1.0	I-23C	

Annex F2. *Continued*

Indicator label	Primary indicator name	Sub-indicator name	Value (0-4)	Weighting factor	Old indicator label	Worksheet location
I-27	Employees		1.9		I-24	Project Employees
I-27A		Frequency and adequacy of training of operators and middle managers (not secretaries and drivers). This should include employees at all levels of the distribution system, not only those who work in the office	2.0	1.0	I-24A	
I-27B		Availability of written performance rules	1.0	1.0	I-24B	
I-27C		Power of employees to make decisions	3.0	2.5	I-24C	
I-27D		Ability of the project to dismiss employees with cause.	2.0	2.0	I-24D	
I-27E		Rewards for ememplary service	2.0	1.0	I-24E	
I-27F		Relative salary of an operator compared to a day laborer	1.0	2.0	I-24F	
I-28		Water User Associations	2.2		I-25	WUA
I-28A		Percentage of all project users who have a functional, formal unit that participates in water distribution	4.0	2.5	I-25A	
I-28B		Actual ability of the strong Water User Associations to influence real-time water deliveries to the WUA	2.0	1.0	I-25B	

Modernisation strategy for the national irrigation systems in the Philippines

Annex F2. *Continued*

Indicator label	Primary indicator name	Sub-indicator name	Value (0-4)	Weighting factor	Old indicator label	Worksheet location
I-28C		Ability of the WUA to rely on effective outside help for enforcement of its rules	0.0	1.0	I-25C	
I-28D		Legal basis for the WUAs	1.0	1.0	I-25D	
I-28E		Financial strength of WUAS	1.0	1.0	I-25E	
I-29	Mobility and Size of Operations Staff	Operation staff mobility and efficiency, based on the ratio of operating staff to the number of turnouts	0.0		I-28	Project office questions
I-30	Computers for billing and record management	The extent to which computers are used for billing and record management	0.0		I-30	Project office questions
I-31	Computers for canal control	The extent to which computers (either central or on-site) are used for canal control	0.0		I-31	Project office questions

Annex F2. *Continued*

Indicator label	Primary indicator name	Sub-indicator name	Value (0-4)	Weight ing factor	Old indicator label	Worksheet location
	INDICATORS THAT WERE NOT PREVIOUSLY COMPUTED	**THESE INDICATORS REQUIRE THE INPUT OF VALUES (0-4) IN EACH OF THE BOXES**				
I-32	Ability of the present water delivery service to individual fields, to support pressurized irrigation methods		0.0		I-26	n/a
I-32A	Measurement and control of volumes to the field	4 - Excellent volumetric metering and control; 3.5 - Ability to measure flow rates reasonably well, but not volume. Flow is well controlled; 2.5 - Cannot measure flow, but can control flow rates well; 0 - Cannot control flow rate, even though it can be measured	0.0	1.0	I-26A	n/a
I-32B	Flexibility to the field	4 - Arranged delivery, with frequency, rate and duration promised. All can be varied upon request; 3 - Same as 4, but cannot vary the duration; 2 - 2 variables are fixed, but arranged schedule; 0 - Rotation	0.0	1.0	I-26B	n/a

Modernisation strategy for the national irrigation systems in the Philippines

Annex F2. *Continued*

Indicator label	Primary indicator name	Sub-indicator name	Value (0-4)	Weighting factor	Old indicator label	Worksheet location
I-32C	Reliability to the field	4 - Water always arrives as promised, including the appropriate volume; 3 - A few days of delay occasionally occur, but water is still very reliable in rate and duration; 0 - More than a few days delay.	**0.0**	1.0	*I-26C*	n/a
I-33	Changes required to be able to support pressurized irrigation methods		**0.0**		*I-27*	n/a
I-33A	Procedures, Management	4 - No changes in water ordering, staff training, or mobility; 3.5 - Improved training, only. The basic procedures/conditions are just fine, they just are not being implemented to their full extent; 3.0 - Minor changes in water ordering, mobility, training, incentive programs; 2.0 - Major changes in 1 of the above; 1 - Major changes in 2 of the above; 0 - Need to completely revamp or convert almost everything	**0.0**	1.0	*I-27A*	Management

Annex F2. *Continued*

Indicator label	Primary indicator name	Sub-indicator name	Value (0-4)	Weighting factor	Old indicator label	Worksheet location
I-33B	Hardware	4 - No changes needed; 3.5 - Only need to repair some of the existing structures so that they are workable again.; 3.0 - Improved communications, repair of some existing structures, and a few key new structures (less than US$300/ha needed), OR…very little change to existing, but new structures are needed for water recirculation; 2 - Larger capital expenditures - US$ 300 - US$ 600/ha; 1 - Larger capital expenditures needed (up to US$ 1500/ha); 0 - Almost complete reworking of the system is needed	**0.0**	1.0	*I-27B*	Hardware
I-34	Sophistication in receiving and using feedback information. This does not need to be automatic.	4 - Continuous feedback and use of information to change inflows, with all key points monitored. Or, minimal feed back is necessary, such as with closed pipe systems.; 3 - Feedback several times a day and rapid use (within a few hours) of that information, at major points.; 2 - Feedback once/day from key points and appropriate use of information within a day; 1 - Weekly feedback and appropriate usage, or once/day feeback but poor usage of the information; 0 - No meaningful feedback, or else there is a lot of feedback but no usage	**0.0**		*I-29*	n/a

Modernisation strategy for the national irrigation systems in the Philippines

Annex F2. *Continued*

Indicator label	Primary indicator name	Sub-indicator name	Value (0-4)	Weighting factor	Old indicator label	Worksheet location
		SPECIAL INDICATORS THAT DO NOT HAVE A 0-4 RATING SCALE				
I-35	Turnout density	Number of water users downstream of employee-operated turnouts	30			*Final deliveries*
I-36	Turnouts/Operator	(Number of turnouts operated by paid employees)/(Paid Employees)	2.5			*Project Office*
I-37	Main canal chaos	(Actual/Stated) Overall Service by the main canal	0.41			
I-38	Second level chaos	(Actual/Stated) Overall Service at the most downstream point operated by a paid employee	0.36			
I-39	Field level chaos	(Actual/Stated) Overall Service to the Individual Ownership Units	0.41			

Annex G. Water estimation techniques used in the sample NIS

Irrigation system, design service area	Drainage area at diversion point	Climate type	Water source	Source runoff data	Period runoff data	Run off generation method	Generated data	Flow availability and dependability
1. Addalam River IP; Quirino and Isabela; New 5,000 ha Rehab 830 ha MARIIS	862 km^2	III	Ungauged	Addalam River gauging station downstream, 896 km^2	Monthly	Transposed mean 10-day @damsite by drainage area proportion	Mean 10-day, MCM	10-day flow duration analysis
2. Macalelon Small Reservoir IP; Macalelon, Quezon;	36.2 km^2	II	Ungauged	Ibia River, 15 km^2 Dumacaa River, 54, km^2	1956-59, 1962-70; 1946-1972 Mean daily Q, m^{3}s^{-1}	Mean dimensionless hydrograph Crawford model (mean daily Q on decadal basis)	Decadal mean 10-day flow, m3^3 s^{-1}	Reservoir operation simulation
3. Quipot River IP; Tiaong, Quezon; 2,800 ha WS/DS	270 km^2	III	Ungauged	Lagnas River, 50 km^2	1952-69; 1971; 1983-99 Monthly Q, m^{3}s^{-1}	Transported @damsite, then Thomas Feiring stochastic model	Monthly flows for 50 yrs, m^{3}s^{-1}	10-day flows frequency analysis; 80% probability

Modernisation strategy for the national irrigation systems in the Philippines

Annex G. Continued

Irrigation system, design service area	Drainage area at diversion point	Climate type	Water source	Source runoff data	Period runoff data	Run off generation method	Generated data	Flow availability and dependability
4. Asbang Small Reservoir IP; Matanao, Davao del Sur; 1660 ha WS/DS	44.7 km^2	IV	Ungauged	Mal River, 188 km^2	1956-78 Monthly runoff, m^{3}s$^-$[1]	Rainfall-runoff regression / correlation analysis using rainfall Davao station (1949-2010) to augment Mal rainfall. Rainfall-runoff regression / correlation analysis using concurrent Mal River gauging record and regressed Mal station rainfall to derive Mal runoff depth base data series, which was converted in m^{3}s^{-1} to derive the runoff for Latian River	Mean monthly flows, m^{3}s^{-1} and MCM	Reservoir water balance study Reservoir operation simulation
5. IBato-Iraan Small Reservoir IP; Aborlan, Palawan; 1062 ha WS 1050 ha DS	23.3 km^2 27.3 km^2	III	Ungauged	Marangas River, 38.4 km^2	Per Main report shown in Appendix 3.2	Runoff-rainfall regression analysis using average monthly rainfall in Aborlan and assessed Marangas River to derive daily Q, which was correlated to Iraan and Ibato Rivers.	Daily flows	Reservoir operation study, monthly 70% dependable flow derived by area correlation method

Water estimation techniques used in the sample NIS

Annex G. *Continued*

Irrigation system, design service area	Drainage area at diversion point	Climate type	Water source	Source runoff data	Period runoff data	Run off generation method	Generated data	Flow availability and dependability
6. Bulo Small Reservoir IP; San Miguel, Bulacan; 570 ha WS 370 ha DS	45.34 km²	I, III	Ungauged	Bulo River gauging station downstream, 57 km²	1964-1976	Rainfall-runoff regression / correlation analysis using concurrent montly data for Bulo rainfall and Bulo River historical runoff at gauging station. Regression equation was used to derive a runoff depth base data series using the whole adapted areal monthly rainfall. Derived runoff was converted in m³s⁻¹ to derive runoff for Dulo at damsite. Drainage coeff is 0.48	Monthly runoff depth, mm Mean monthly runoff, m³s⁻¹ and MCM	Reservoir operation study,

Modernisation strategy for the national irrigation systems in the Philippines

Annex G. *Continued*

Irrigation system, design service area	Drainage area at diversion point	Climate type	Water source	Source runoff data	Period runoff data	Run off generation method	Generated data	Flow availability and dependability
7. Balbalun gao Small River IP; Lupao, Nueva Ecija; 840 ha WS, 640 ha DS	9.67 km²	I	Ungauged	Talavera River gauging station, 261 km²	1956-74 (w/ gaps)	Rainfall-runoff regression/correlation analysis bet. Pantabangan augmented rainfall (X variable)and Talavera River runoff (Y variable) using concurrent monthly data. Used the resulting regression equation to derive runoff depth base data series using the whole adapted monthly rainfall of Pantabangan. Converted the derived runoff depth in m^3s^{-1} units to derive runoff for Balbalungao Rivera @damsite.	Mean monthly runoff, m^3s^{-1} and MCM, 1965-2009	Reservoir water balance study or reservoir behavioral analysis

Water estimation techniques used in the sample NIS

Annex G. *Continued*

Irrigation system, design service area	Drainage area at diversion point	Climate type	Water source	Source runoff data	Period runoff data	Run off generation method	Generated data	Flow availability and dependability
8. Balog-balog Multi Purpose Proj; San Jose, Tarlac; New area 21,935 ha WS/DS	283 km^2	I	Ungauged	Lower Bulsa River, O' Donnel River Camiling River	None shown	Used available short record of streamflow from lower Bulsa, O'Donnel and Camiling rivers to derive the regional characteristics of the upper Bulsa River basin in terms of specific low flow and approximate dry flow depletion rate. Use water balance technique and 30-yr daily rainfall data, computed daily PET and contribution of base flow component to derive long-term discharge data series. Used deterministic rainfall-runoff hydrologic model to generate stream flow for the Upper Bulsa River using daily rainfall and computed daily PET.	Mean monthly flows at damsite, m^{3}s^{-1}	Flow duration curve of generated flows at damsite Reservoir operation simulation

Modernisation strategy for the national irrigation systems in the Philippines

Annex G. *Continued*

Irrigation system, design service area	Drainage area at diversion point	Climate type	Water source	Source runoff data	Period runoff data	Run off generation method	Generated data	Flow availability and dependability
8. *Continued*						Parameters for base flow and basin characteristics were derived from contour maps and characteristics of low flow of similar rivers in the region. These parameters where used in calibrating the hydrologic model for generating daily flows. Estimated the time lag of surface runoff from rainfall occurrence based on catchment area, hydrologic gradient, estimated river basin curve number CN. Assessed validity of simulation results by comparing specific annual runoff of the generated flow with the average regional specific annual runoff from available regional water resources data of NWRB. The mean specific monthly flow of the short flow records were compared the generated flows		

Water estimation techniques used in the sample NIS

Annex G. *Continued*

Irrigation system, design service area	Drainage area at diversion point	Climate type	Water source	Source runoff data	Period runoff data	Run off generation method	Generated data	Flow availability and dependability
9. Kitcharo SRIP; Kitcharao, Agusan Del Norte; 550 ha, WS/DS	5.20 km²	II	Mankas River; ungauged	Butuan rainfall data	1956-1989 monthly rainfall	Conversion of rainfall data to composite hydrograph of flow by using linear time-in variant rainfall-runoff system which considers the principle of linearity and superposition. Developed a synthetic hydrograph for the river basing by suing topomap and present condition of the river basin. Approximated the rainfall over Kitcharao River Basin by applying inverse distance method on Butuan and Surigao rainfall stations. The derived rainfall for Kitcharao River Basin was converted to effective rainfall by subtracting the losses. Then, input to rainfall-runoff method.	1956-1989 Monthly runoff for Mankas River, MCM	Reservoir operation study

Modernisation strategy for the national irrigation systems in the Philippines

Annex G. *Continued*

Irrigation system, design service area	Drainage area at diversion point	Climate type	Water source	Source runoff data	Period runoff data	Run off generation method	Generated data	Flow availability and dependability
10. Libmanan-Cabusao Dam Project; Camarines Sur; 4,000 ha	442.1 km²	II	Sipocot River	Sipocot River gauging station	1969-1984	Derived runoff at the proposed dam site from the runoff of Sipocot River at the gauging station by using drainage area proportion ration method. Filled and extended the runoff data by using rainfall-runoff correlation between Baao rainfall data and Sipocot River discharges from 1969-1984.	10-day and monthly runoff at gauging station; derived 10-day and monthly runoff at dam site;	Used flow duration analysis to compute 80% probability for every decadal
11. Pasa SRIP; Isabela; 800 WS/DS	18.33 km²	III	Pasa River; ungauged	Pinacanauan de Tumaini River gauging station (DA=170 km²)	1964-1970	Rainfall-runoff regression/correlation analysis between concurrent historical monthly CVIARC rainfall data and Pinacanauan de Tumauini runoff	1961-2010 Mean monthly runoff data at dam site, in CMS and MCM	Reservoir operation study

Water estimation techniques used in the sample NIS

Annex G. *Continued*

Irrigation system, design service area	Drainage area at diversion point	Climate type	Water source	Source runoff data	Period runoff data	Run off generation method	Generated data	Flow availability and dependability
12. Jalaur RIS; Iloilo; 34,340 ha	106.97 km², 72.61 km²	I	Jalaur River, Ulian River	Jalaur River at Alibunan Calinog (DA = 120 km²)	1956-1970	Rainfall-runoff hydrologic model was used to generate streamflows for Jalaur and Ulian Rivers using the 1975-2004 years daily rainfall from Iloilo Synoptic Station.	1975-2004 Daily & mean monthly flows; flow duration curves	Comparison of annual runoff ratio of generated streamflows to the regional and/or provincial values
				Ulian River at Pader Dueñas (DA = 247 km²)	1960-1970	Parameters on base flows and the basin characteristics of the two rivers were derived from the 1956-1970 streamflow records and, then used in calibrating the hydrologic model for generating streamflows on daily basis.		

Annex G. *Continued*

Irrigation system, design service area	Drainage area at diversion point	Climate type	Water source	Source runoff data	Period runoff data	Run off generation method	Generated data	Flow availability and dependability
13. Casecnan Multipurpose IP	520 km^2	I	Casecnan (Abaca) and Denip Rivers	Gibong gauging station (in Casecnan basin); DA= 832 km^2	1977-1982	Used rainfall runoff model to estimate long term streamflow from rainfall after calibration with observed streamflow data. Tank model is used to generate daily discharges. Used the 1974-1982 rainfall records at Casenan and Dakgang stations to determine the basin rainfall using isohyetal method. The tank coefficients were determined from 1977-1982 river discharges and estimated basin rainfall by trial and error. These coefficients were used to generate streamflows at Casecnan and Denip dams	1949-1999 Mean monthly runoff, mean annual river discharge, runoff	

Water estimation techniques used in the sample NIS

Annex G. *Continued*

Irrigation system, design service area	Drainage area at diversion point	Climate type	Water source	Source runoff data	Period runoff data	Run off generation method	Generated data	Flow availability and dependability
14. Sta. Josefa Pump IP; Agusan del Sur; 2,000 ha	1,540 km²	II	Agusan River	Agusan River; DA = 1,359 km²	1978-2002 Daily, monthly	Runoff-drainage area proportion	10-day runoff, 1978-2002	80% probability
15. Chico River Pump IP; Kalinga and Cagayan Valley; 8,700 ha	3,355 km²	III	Chico River	Saltan River	1965-1972	Rainfall-runoff correlation between rainfall data for Tuguegarao and Cagumitan stations and Saltan River flow data	10-day mean monthly runoff, 1967-2010	R = 75.4%; probability of 80%

Annex H. List of streamflow gauging stations in Region IV-A

	River name/Station ID	Location	POR[1]	Drainage area[2]	Status[3]
1	Balay-Balay, Baler	Mauban, Quezon	1985	n.d.	O
2	Dacanlao River	Balayan, Batangas	1982	n.d.	O
3	Dumacaa 04SW140213BRS033	Alsam, Tayabas, Quezon	1983-1985	54	A
4	Ilang-ilang River	Alapan II, Imus, Cavite	1982-1985	n.d.	O
5	Iyam River	Lucena City, Quezon	1987	n.d.	O
6	Laguna Lake (Tidal) 04SW141211PW066	Los Baños, Laguna	1984-2003	3,158	O
7	Maapon River	Sampaloc, Quezon	1983	n.d.	O
8	Maragondon 04SW141204PW068	Brgy. Bukal, Maragondon, Cavite	1983-2005	242	O
9	Mayor 04SW142212BRS069	Famy, Laguna	1983-1993; 1994-2007	48	O
10	Pagsanjan River	San Isidro, Pagsanjan, Laguna	1984-1999	n.d.	O
11	Palico 04SW140204BRS072	Palico, Nasugbu, Batangas	1985-2000	167	A
12	Panaysayan 04SW142205PW023	Palubluban, General Trias, Cavite	1983-2001; 2002-2004	30	O
13	Pansipit 04SW135205PW029	Poblacion, San Nicolas, Batangas	1983-2007	673	O
14	Pililia 04SW142211BRS073	San Lorenzo, Pililia, Rizal	1985-1992	26	O
15	San Cristobal 04SW141210BRS074	San Cristobal, Calamba, Laguna	1984-1999	106	A
16	San Juan River	Tanauan, Batangas	1986-1999	n.d.	O
17	San Juan River	Porac, Calamba, Laguna	1986-1999	n.d.	O
18	San Roque 04SW134205BRS076	San Roque, Bauan, Batangas	1987-2002	16	A
19	Sariaya 04SW135213PW036	Tumbaga, Sariaya, Quezon	1983-1993	4	O
20	Sta. Cruz River	Sta. Cruz, Laguna	1985	n.d.	O
21	Taal Lake	San Nicolas, Batangas	2013	n.d.	O
22	Tignoan 04SW143213BRS078	Tignoan, Real, Quezon	1985-2007	85	O

[1] with processed data for the given POR; a given single year mean start of record and with unprocessed data; [2] n.d. - no data; [3] O - operational station, A - abandoned station

Annex I. Assessment of relative quality of water deliver service and demand for canal operation

Quality of water delivery service: definitions and assigned points

Equity vis-à-vis most upstream TSA	
Description:	*Rate*
Equal	4
Almost equal	3
Low	2
Lower	1
> very low	0
Flexibility	per RAP definition
Reliability	per RAP definition
Volume measurement	per RAP definition

Demand for canal operation: definitions and assigned points

Crop risk to drought or flood	No	Yes	Limited
Water shortage in any cropping season	0	1	0.5
Without supplemental canal water	0	1	
Flood-prone area	0	1	
Sensitivity: Hydraulic type of diversion structures			
Ungated, non-functional gates	3		
Functional structures with adjustable gates	2		
Functional, fixed proportional structures	1		
Perturbation: Percentage of ungated offtake/inlet			
1/3 of ungated offtake/inlet upstream	1		
2/3 of ungated offtake/inlet upstream	2		
3/3 of ungated offtake/inlet upstream	3		

Demand for canal operation: Ranges of total point and demands ratings

Ranges of total points	Demand category (1 is lowest, 5 highest demand)
0-2	1
>2-4	2
>4-6	3
>6-8	4
>8-10	5

Annex I. *Continued*

Relative quality of water delivery service in Balanac RIS

TSA	Water delivery service indicators and relative weight factors				Quality indicator	Mapped category
	Volume measurement	Flexibility	Reliability	Equity		
	1	2	4	4		
1	0	4	2	4	2.9	3
2	0	4	2	4	2.9	3
3	0	4	2	4	2.9	3
4	0	4	2	4	2.9	3
5	0	4	2	4	2.9	3
6	0	4	2	4	2.9	3
7	0	4	2	3	2.5	3
8	0	4	2	4	2.9	3
9	0	4	2	4	2.9	3
10	0	4	2	1	1.8	2
11A	0	2	1	2	1.5	1
11B	0	2	1	1	1.1	1
12	0	2	1	1	1.1	1
13	0	2	1	1	1.1	1
14	0	2	1	1	1.1	1
Buboy 1	0	2	2	2	1.8	2
Buboy 2	0	2	2	4	2.5	3
Buboy 3	0	2	2	2	1.8	2
Buboy 4	0	2	2	2	1.8	2
Buboy 5	0	2	2	2	1.8	2
Lat A-1	0	2	2	2	1.8	2
Lat A-2	0	2	2	2	1.8	2
Lat A-3	0	2	2	4	2.5	3
Lat A- 4	0	2	2	4	2.5	3
Lat A1-1	0	2	2	4	2.5	3
Lat A1-2	0	2	2	1	1.5	1
Lat A1A-1	0	1	1	1	0.9	1
Lat A1A-2	0	1	1	1	0.9	1
Lat A1A-3	0	1	1	2	1.3	1
Biñan 1A	0	2	2	2	1.8	2
Biñan 1B	0	2	2	2	1.8	2
Biñan 2	0	1	1	2	1.3	1
Salasad 1	0	2	2	4	2.5	3
Salasad 2	0	2	2	2	1.8	2

Annex I. *Continued*

Relative quality of water delivery service in Sta. Maria RIS

TSA	Water delivery service indicators and relative weight factors				Quality indicator	Mapped category
	Volume measurement	Flexibility	Reliability	Equity		
	1	2	4	4		
1	0	2	2	4	2.5	3
2	0	2	2	3	2.2	2
3	0	2	2	3	2.2	2
4	0	2	1	3	1.8	2
5	0	2	1	2	1.5	1
5A	0	2	1	2	1.5	1
6	0	2	1	2	1.5	1
6A	0	2	1	2	1.5	1
7	0	2	1	2	1.5	1
8	0	2	1	2	1.5	1
9	0	2	1	3	1.8	2
10	0	2	1	3	1.8	2
11	0	2	1	2	1.5	1
12	0	2	1	2	1.5	1
13	0	2	1	2	1.5	1
14	0	2	1	2	1.5	1
15	0	2	0	1	0.7	1
16	0	2	1	4	2.2	2
17	0	2	1	4	2.2	2
18	0	2	2	4	2.5	3
19	0	2	2	4	2.5	3
20	0	2	0	3	1.5	1
21	0	2	2	4	2.5	3
22	0	2	2	4	2.5	3
23	0	2	2	4	2.5	3
24	0	2	2	4	2.5	3
25	0	2	2	3	2.2	2
26	0	2	0	3	1.5	1
27	0	2	0	2	1.1	1
28	0	2	0	2	1.1	1

Annex I. *Continued*

Relative demand for Balanac RIS canal operation as defined by demand for service, sensitivity and perturbations

TSA	Demand for service	Sensitivity	Perturbations	Demand for canal operation	Demand category
1	0	3	1	4	2
2	0	3	1	4	2
3	0	3	1	4	2
4	0	3	1	4	2
5	0	3	1	4	2
6	0	3	1	4	2
7	0	3	1	4	2
8	0	3	2	5	3
9	0	3	2	5	3
10	1	3	2	6	3
11A	2.5	3	2	7.5	4
11B	1.5	3	2	6.5	4
12	2.5	3	3	8.5	5
13	3	3	3	9	5
14	3	3	3	9	5
Buboy 1	0	3	1	4	2
Buboy 2	0	3	1	4	2
Buboy 3	0	3	1	4	2
Buboy 4	0	3	1	4	2
Buboy 5	0	3	2	5	3
Lat A-1	0	3	2	5	3
Lat A-2	1.5	3	2	6.5	4
Lat A-3	0	3	2	5	3
Lat A-4	0	3	2	5	3
Lat A1-1	2	3	2	7	4
Lat A1-2	2	3	2	7	4
Lat A1A-1	2	3	3	8	4
Lat A1A-2	2.5	3	3	8.5	5
Lat A1A-3	1.5	3	3	7.5	4
Biñan 1	0	3	1	4.0	2
Biñan 2	1	3	1	5.0	3
Biñan 3	3	3	1	7	4
Salasad 1	1	3	1	5	3
Salasad 2	1	3	1	5	3

Annex I. *Continued*

Relative demand for Sta. Maria RIS canal operation as defined by demand for service, sensitivity and perturbations

TSA	Demand for service	Sensitivity	Perturbations	Demand for canal operation	Mapped category
1	0	2	1	3	2
2	1	2	1	4	2
3	1	2	1	4	2
4	1	2	1	4	2
5	2	2	1	5	3
5A	1.8	2	2	5.8	3
6	1.5	2	2	5.5	3
6A	2.5	2	2	6.5	4
7	2	2	1	5	3
8	2	2	1	5	3
9	1.1	2	1	4.1	2
10	1.1	2	1	4.1	2
11	2	2	1	5	3
12	2	2	1	5	3
13	2	2	1	5	3
14	2	2	2	6	3
15	2	2	2	6	3
16	1	2	1	4	2
17	1	2	1	4	2
18	0.8	2	1	3.8	2
19	0.8	2	1	3.8	2
20	1	2	2	5	3
21	1	2	2	5	3
22	2	2	1	5	3
23	0	2	2	4	2
24	0	2	2	4	2
25	1	2	2	5	3
26	1	2	3	6	3
27	3	2	3	8	4
28	3	2	3	8	4

Annex J. The drainage network in the service area of Sta. Maria RIS

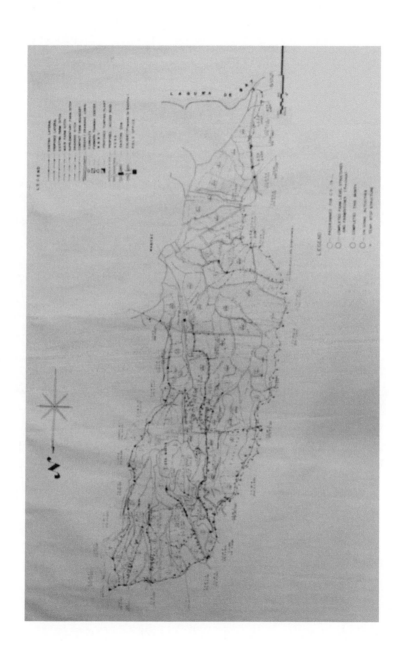

Annex K. Front line O&M expenses of Balanac RIS and Sta. Maria RIS (PhP 10³)

Balanac RIS	2010	2011	2012	2013	2014	2015	Total
Personnel			942.4	991.2	928.5	1,243.4	4,105.5
Wages			836.9	907.8	881.8	1,195.1	3,821.5
SSS premium			-	-	-	-	-
Honorarium			88.1	72.0	46.7	46.4	253.1
Incentive			-	-	-	-	-
Training			17.5	11.4	-	1.9	30.9
Repairs & maintenance			94.4	59.5	93.4	92.9	340.1
Riprap			-	-	-	-	-
Canal clearing			94.4	59.5	93.4	92.9	340.1
Travel			112.2	87.6	34.0	33.9	267.7
Fuel/oil			89.5	47.4	24.7	24.6	186.2
Hired transport			-	-	-	-	-
Official travel			22.7	40.1	9.3	9.3	81.5
Tax, license			-	-	-	-	-
Office administration			94.9	84.9	20.6	20.5	221.0
Office supplies			44.7	1.8	2.0	2.0	50.5
Electric bill			35.8	35.1	18.7	18.6	108.1
Photocopying			8.4	42.1	-	-	50.5
Admin repair & maintenance			-	-	-	-	-
Registration			6.0	5.9	-	-	11.9
Miscellaneous			49.3	64.8	184.8	828.1	1,127
Representation allowance			-	-	-	-	-
Meetings -meals/snacks			13.5	5.8	1.8	1.9	23.0
Donations/gifts			6.8	3.0	5.6	5.7	21.1
Christmas			-	-	-	-	-
Sundries			28.9	55.9	177.4	176.5	438.8
Non-personnnel expenses			-	-	-	644.1	644.1

Annex K. *Continued*

Sta. Maria RIS

Personnel	523.6	619.6	789.9	896.8	831.2	914.4	4,575.6
Wages	435.1	512.3	639.8	711.6	706.9	808.7	3,814.4
SSS premium	15.9	20.2	18.6	21.1	21.7	26.0	123.5
Honorarium	72.6	82.2	90.8	111.7	91.6	73.5	522.4
Incentive	-	1.9	34.2	41.6	7.5	-	85.3
Training	-	3.0	6.5	10.8	3.5	6.2	30.0
Repairs/maintenance	145.9	158.2	230.5	100.6	78.9	83.3	797.5
Riprap, repair, maintenance	-	25.0	10.8	4.6	8.2	83.3	131.9
Canal clearing	145.9	133.1	219.7	96.1	70.8	-	665.5
Travel	99.9	121.3	92.7	112.1	101.6	82.3	609.8
Fuel/oil	89.3	114.7	87.2	95.0	83.7	72.7	542.5
Hired transport	3.6	3.4	4.7	11.1	8.6	7.5	38.7
Official travel	7.0	3.1	0.8	6.0	9.4	2.1	28.6
Office administration	7.7	23.7	24.4	73.1	35.4	47.9	212.1
Office supplies	1.2	13.6	5.5	24.0	5.9	14.2	64.4
Electric bill	3.3	3.9	3.6	3.6	5.0	4.3	23.8
Photocopying	1.0	0.7	1.0	1.4	1.2	0.5	6.0
Admin repair & maintenance	-	-	7.6	15.3	1.1	-	24.0
Tax, license	2.1	5.6	6.7	20.7	3.5	9.1	47.6
Depreciation	-	-	-	7.4	18.5	19.7	45.5
Notarial fee	-	-	-	0.7	0.2	-	0.9
Miscellaneous	18.2	40.6	33.8	44.6	20.3	10.8	168.3
Representation allowance	9.1	0.4	1.5	-	-	-	11.0
Meetings -meal/snacks	1.2	13.0	13.0	10.4	15.6	7.1	60.3
Donations/gifts	-	14.8	12.6	2.5	-	-	29.9
Christmas	5.3	11.9	-	-	-	-	17.2
Sundries	2.6	0.5	6.7	31.7	4.8	3.8	50.0
Non-personnnel expenses	-	-	-	-	-	-	-

Annex L. Groundwater map of Laguna

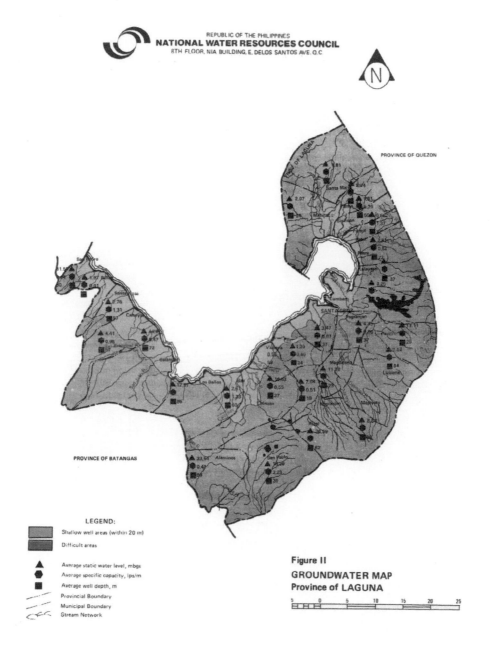

Figure II

GROUNDWATER MAP
Province of LAGUNA

Annex M. Well drilling in Sta. Maria RIS service area

Drilling site 1: 14.50835° N, 121.42311° E; Adjacent to Sta. Marial main canal in TSA 13

Drilling site 2: 14.44339° N, 121.41962° E; Adjacent to TSA 28

Annex N. Vision for the case study systems

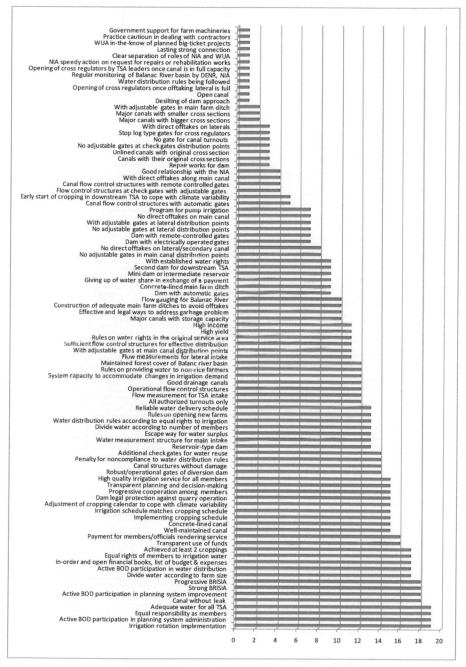

Details of the water users' vision for Balanac RIS

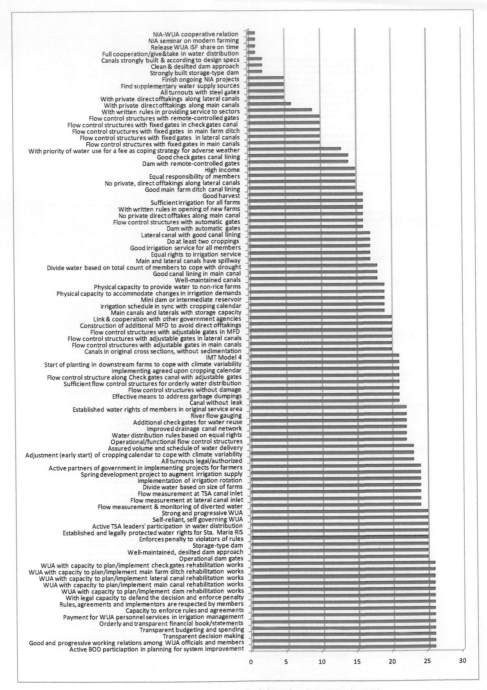

Details of the water users' vision for Sta. Maria RIS

Annex O. Summary

The actual areas irrigated by publicly funded canal irrigation systems, commonly known as national irrigation systems (NIS) in the Philippines, are significantly lower than their design service areas. The associated cropping intensities and crop yields are below the projected levels during the design stage. Such mediocre performance has been attributed to design shortcomings related to the use of too optimistic assumptions on canal hydraulics, water supply availability and water demand parameters such as seepage, percolation and evapotranspiration.

In most cases, the current design features of physical structures are not amenable to efficient and effective management and distribution of irrigation water. These include lack of head control structures, hydraulically inconsistent combinations of flow control structures, direct offtakes along main canals, ungated offtake structures, insufficient discharge capacity of canals and offtakes, and inadequate protection of main intake structures from siltation. Also, a number of studies pointed to overestimation of surface water supply availability and underestimation of seepage and percolation as causes of a low ratio of irrigated area to irrigation service area.

The government on its part has been implementing various irrigation system rehabilitation and improvement projects. Despite considerable rehabilitation and improvement efforts, about 30-40% of the irrigation service areas of the country are not irrigated. The changes in weather pattern, water supply availability, crops, irrigation water and service requirements and physical structures of irrigation systems add some difficulty to irrigating the whole service areas.

Modernisation of irrigation systems is now viewed as a process of technical and managerial upgrading to improve irrigation service to farmers. This becoming widely accepted definition of modernisation distinguishes it from rehabilitation, which simply restores the physical design of the original system. With an increasing competition for water and land and increasing cost of developing new irrigation systems, experts say the performance of existing canal irrigation systems in terms of water use and crop production has to improve in order to increase the chance of meeting the global food requirement in the near future. They hold the opinion that irrigation system modernisation is a strategic option to improve the performance of canal irrigation systems. As many of the irrigation systems have aged, weather patterns and hydrographs of water sources have changed and farmers' desire to intensify cropping has increased, the physical structure of existing irrigation systems and the associated canal operation requirements have become less responsive to the present irrigation service demands. Thus, upgrading of irrigation structures and system operations has become a necessary undertaking.

The general objective of this research was to identify technically feasible and hydrologically sound modernisation plans for national irrigation systems in the Philippines. To achieve this objective, the following studies have been carried out and the findings synthesized: (1) analysis of the nature of previous rehabilitation or modernisation efforts and their impacts in terms of closing the gap between the actual area irrigated and the design and firmed-up service areas; (2) assessment of the efficacy of the physical structure and operation of irrigation systems to deliver the design rate, duration and frequency of irrigation water or a modified irrigation water demand at the offtakes of secondary canals; (3) assessment of the validity of design values for dependable water supply, crop water requirement, seepage and percolation; (4) identification of modernisation options for the realistic NIS service areas; (5) investigation of the potentials of shallow tubewell irrigation in the part of NIS service areas that cannot be fully irrigated with surface water as designed; (6) identification of possible options of integrating conjunctive use of groundwater and surface water in the formulation of a modernisation plan for NIS.

Three NIS located in Region IV-A of the country were selected for the case study, namely: the Agos River Irrigation System (RIS), Balanac RIS and Sta. Maria RIS. The selection was based on a classification of the condition of the physical facilities as a result of previous rehabilitation works, hence, readiness for irrigation management transfer (IMT). Under such classification, Balanac RIS and Sta. Maria RIS are considered ready for IMT after minor repairs of some of their physical components. Meanwhile, Agos RIS is deemed in need of major rehabilitation works before it becomes amenable for operation, maintenance and management by farmers.

The 1980-2012 data on rehabilitated, restored and newly generated service areas vis-a-vis the cumulative service areas and actual irrigated areas nationwide were analyzed to assess the impacts of rehabilitation in terms of maintaining developed irrigation service areas and closing the gap between service areas and actual areas irrigated. Similar trend analysis was carried out for the 1990-2012 data for each of the three systems. The nature of rehabilitation works or the physical components involved in the rehabilitation projects implemented for each of the systems since mid-1990s were examined. Key persons at different levels of NIA offices and concerned water users associations (WUA) were interviewed to solicit information and insights on the planning and implementation process of rehabilitation projects.

From 1980 to 2012, the total service area of NIS in the country increased by almost 319,000 ha. During the same period, the new service area generated by constructing new NIS and expanding the service areas of the existing systems beyond the original size was about 415,000 ha. A total of about 3.5 million ha NIS service areas were rehabilitated during this period. This implies that the country's NIS service area was rehabilitated once every 6-7 years since 1980. On the average, the rehabilitated NIS area was about 1.6 times its service area for each decade. Meanwhile, the average annual rates of rehabilitation for the systems under study were about 75% of the firmed-up service area (FUSA) of Balanac RIS and 40% of FUSA of Agos RIS and Sta. Maria RIS. The three systems have undergone more frequent rehabilitation than that of the national level.

At the country level, rehabilitation projects were neither able to maintain the irrigation service area nor to close the gap between the irrigation service area and actual irrigated area. In the case of the three systems, rehabilitation had, at best, maintained the percentages of the FUSA actually irrigated since 2000. The magnitude of rehabilitation reflected a fast rate of deterioration of the irrigation systems in the country. These findings point some questions with respect to the efficiency of planning and implementation of canal irrigation projects. It casts doubt on the effectiveness of the rehabilitation efforts.

The findings of the analysis of underlying details of rehabilitation projects show that the rehabilitation works generally focused on restoring the original physical structure. Concrete canal lining was the most frequent and most invested activity, accounting for about 50-60% of the total rehabilitation expenses for each NIS. Headworks, roads and canal structures were among the next two most invested projects in the three NIS. They each comprised less than 10% of the total investment for the respective systems with the exception of the dam of Balanac RIS and road in Sta. Maria RIS, which accounted for 22 and 15%, respectively.

The adopted planning and design process lacked revalidation of design criteria and assumptions used in the original system design. It did not have the benefit of diagnostic assessment studies. No technical assessments of water supply adequacy for proposed projects were carried out. The rehabilitation process lacked an impact evaluation component.

A combination of a logic design framework by Ankum, diagnostic tools of the mapping system and services for canal operation techniques (MASSCOTE), discharge-head relations and hydraulic flexibility were applied to diagnose the performance of Balanac RIS and Sta. Maria RIS and identify options for system improvements. The logic design framework examines the coherence of the design philosophy, overall system objectives, objectives of system operation, design configuration of physical structures and flow control methods. For a set of stated design philosophy and system objectives, certain design parameters, system configuration and combinations of flow control structures fit well or maybe the only logical match. Meanwhile, the diagnostic tools of MASSCOTE, which include the rapid appraisal procedure (RAP) and assessments of physical capacity and hydraulic behaviour (sensitivity and perturbations) of irrigation systems, are aimed at systematic and comprehensive evaluation of the performance of large-scale canal irrigation systems to modernize canal operation. They consider the water balance, internal processes and mechanisms (hardware, operational procedures, management and institutions) of water delivery at different canal levels, irrigated area and crop production in diagnosing system performance.

A review of the general design considerations and guidelines for canal irrigation of the NIA and interviews with system officials on history of system development including previous rehabilitations, changes in system design and management were carried out. The design philosophies, system objectives, design configurations of physical structures and flow control methods were derived from this collected information and were classified according to the parameters of the logic design framework.

The RAP worksheets were used as a set of questionnaires for collecting the information required for system diagnosis. The information was obtained from system documents, the weather agency as well as through system walkthroughs and interviews of irrigation superintendents, operation staff and water users associations. The walkthroughs from the dam to the end checks of main canals and major distribution canals of each NIS were carried out as part of the RAP in getting an impression on the current state of physical structures and system operations. The values of RAP indicators of system performance were calculated from the RAP computer spreadsheet data.

The applicability of the water-related RAP external indicators and structures sensitivity assessment as diagnostic tools was limited by the lack of flow data and the infeasibility of field experiments due to tight rotational irrigation schedules and laissez-faire direct offtakes. Hence, a pragmatic approach of using discharge-head relations, hydraulic flexibility concept, and the findings from walkthroughs and interviews was adopted for the analyses.

The result of categorization of the system profile data based on the logic design framework showed that the original Balanac RIS was designed for 'productive irrigation' during the dry season based on 'equitable supply per hectare'. The operational objective was 'imposed allocation' to tertiary units or irrigation service delivery points by 'adjustable flow' and with 'adjustable flow' through the main system or major conveyance canals and employing upstream control. It was equipped with adjustable vertical gates for controlling offtake discharges and water levels upstream. The original design of Sta. Maria RIS had these same philosophies, overall system objectives and operational objectives, except in the case of the design irrigation season. Its design water duty of $0.8 \ 1 \ s^{-1}ha^{-1}$ is very low and logically suggests that the design 'irrigation season' was the wet season. In other words, the system was designed to supplement rainfall and could not irrigate the whole service area during the dry season.

There existed a logical coherence among the design philosophy or overall system objectives, operational objectives and on-canal flow structures of the original design of Balanac RIS and Sta. Maria RIS. However, the shift to 'splitted' flow and 'proportional control' manifested by duckbill weirs at major bifurcation points of Balanac RIS was not consistent with its unchanged overall system objectives. In contrast, the reduction in service area of Sta. Maria RIS and the shift to the dry season as the main irrigation season had maintained the coherence among its objectives. Further, the addition of open direct turnouts along the main canals of both systems is inconsistent with the systems' operational objectives.

In general, the values of the primary internal indicators of RAP calculated for the two systems were low, ranging from 0 - 2 (0 and 4 indicating least and most desirable, respectively). The only exceptions were in the case of 'travel time of a flow rate change through the main canals' in both systems and in the case of control of the 'cross-regulators along the main canals' of Sta. Maria RIS. The results of RAP carried out in Balanac RIS support the hypothesis of unwieldy water distribution deduced from the logic design framework analysis. The values of RAP internal indicators for water delivery and control and operation of flow structures for the two systems were low mainly due to the lack of functional flow control structures and presence of direct, open turnouts along the main canals and other major conveyance canals.

The capacity of physical structures of Balanac RIS and Sta. Maria RIS to perform their intended functions had decreased due to damage, defects, dysfunctions, missing parts and deviations from preconditions for proper functioning through time. Repairs or replacements are needed for non-functional structures and maintenance works to restore head differentials at the dam and approach conditions for the flumes. In Balanac RIS and Sta. Maria RIS, the most telling capacity issues were the division capacity and the limited water supply from the rivers, respectively. Options to increase the storage capacity of Sta. Maria dams or augmenting the water supply from other sources need to be investigated.

The original diversion structures at major distribution points of Balanac RIS and Sta. Maria RIS consisted of adjustable underflow discharge and water level controls, which are amenable for gate proportional diversion or attaining same discharge variations at major head-end and tail-end offtakes of the systems. However, the required frequent gate adjustments to maintain proportional diversion makes adjustable underflow structures a less practical match for the runoff-off-the-river dams and variable water supply from the rivers. The present lack of functional flow control structures and combinations of overflow and underflow at major distribution points result in unproportional discharge variations at the offtaking canals, continuing canals and other downstream offtakes. The many ungated direct offtakes further contribute to inherent flow variations and perturbations along the major canals.

Field experiments and in-depth review of design values and assumptions on percolation and water supply were carried out to gauge their validity and the accuracy of the projected irrigation service areas. Also, the validity of the estimates was assessed in terms of the percentage of programmed irrigation area actually served. As copies of original feasibility study and design documents no longer existed at NIA offices, design engineers, hydrologists and operation engineers of the NIA were interviewed on the procedures and methods used by the agency in estimating available streamflow for irrigation systems. Further, the procedures and methods adopted in estimating the service areas in feasibility studies for 20 sample irrigation projects implemented during 2010-2014 were reviewed.

The percolation tests using the 3-cylinder method and ponding tests were conducted in 19 sites within the service areas of Balanac RIS and Sta. Maria RIS. A collaborative study to determine the physical and hydraulic characteristics of the soil of the experiment sites was carried out with students. The parameters that were identified included the soil texture, bulk density, particle density, porosity and hydraulic conductivity. They were determined through hydrometer, oven dry, volume displacement and falling-head permeater methods, respectively. The results of soil texture analyses were used to identify the percolation values for each soil type as cited in literature and, eventually, to countercheck the measured percolation values.

The measured percolation rates in Balanac RIS have a wide range of values, averaging from 1-30 mm day^{-1}. The design value of 2 mm day^{-1} for the system was valid in only 3 of the 9 experiment sites. The average percolation rates were greater than 28 mm day^{-1} in two of the nine sites. The respective average percolation in another two sites differed significantly when the conditions of the surrounding soils of the experiment fields changed - from 5 mm day^{-1} with settled soil to 22 mm day^{-1} during newly ploughed and from 12 mm day^{-1} in saturated to less than 1 mm day^{-1} when unsaturated. In Sta. Maria RIS, the measured average percolation rates in the 10 sites ranged from 1-6 mm day^{-1}. The design value of 1.4 mm day^{-1} for the system was observed in five of the 10 experiment sites. The average percolation values of about 6 mm/day and 3-4 mm day^{-1} were obtained in one site and four experiment sites, respectively.

The results of soil texture analyses showed some differences between the soil type in the experiment sites and those indicated in the official soil classification maps, which range from clay to clay loam in Balanac RIS and clay in Sta. Maria RIS. The most notable difference in Balanac RIS was the occurrence of sandy soils in three experiment sites. The percolation rates obtained in sites with sandy soils had 25-30 mm day^{-1}, which are much higher than the 2 mm day^{-1} conventionally used in the design for sandy clay loam soils and 4 mm day^{-1} for sandy loam soils. Clay loam and loam soils were also found in six of the 10 experiment sites in Sta. Maria RIS. For the four sites with predominantly clayey soils, two had measured percolation rates within the general design value of 1.25 mm day^{-1} for clay soils. Meanwhile, majority of the sites with clay loam soils had percolation rates almost twice as much as the conventional 1.75 mm day^{-1} design values for percolation.

Although the relative rates of percolation based on soil texture classification were observed, the actual values of percolation significantly differed from their associated values given in the literature and design guidelines. Other factors that could have influenced the percolation rates were the relative elevation of the experiment site to the adjacent fields, welling up of shallow groundwater and ploughing that loosens or breaks the hard pan. Thus, sample in-situ measurements should be done at least per tertiary service area.

The results of the study suggest that the potentials for increasing the actual area irrigated lie in augmentation of water supply or increasing the system storage capacity, use of flow control structures that are amenable to flexible water distribution schemes and strategic timing of irrigation schedules among tertiary areas to adapt to prevailing hydrological regimes.

The experiences in the three sample NIS showed that a more systematic approach to irrigation system improvements would need to be pursued. This approach would include diagnostic assessment of the physical structure, operation, maintenance and water delivery of irrigation systems; identifying constraints, potentials and options for improvements; ascertaining water adequacy for improvement projects; due consideration to the acceptability of improvement options to farmers; development of a blueprint for modernisation plans for individual systems, and establishing a strategic monitoring and evaluation (M&E) system. The observations, analyses, findings and recommendations as presented in this thesis are expected to create a solid basis for such activities.

Annex P. Samenvatting

De werkelijk geïrrigeerde gebieden met door de overheid gefinancierde kanaal irrigatiesystemen, in de Filippijnen beter bekend als de nationale irrigatiesystemen (NIS), zijn aanzienlijk geringer dan de ontworpen gebieden. De bijbehorende gewas intensiteiten en gewasopbrengsten zijn minder dan de tijdens de ontwerpfase geraamde niveaus. Een dergelijk middelmatig functioneren wordt toegeschreven aan tekortkomingen in het ontwerp met betrekking tot het hanteren van te optimistische veronderstellingen betreffende kanaal hydraulica, de beschikbaarheid van water en parameters betreffende de waterbehoefte zoals kwel, percolatie en gewasverdamping.

In de meeste gevallen zijn de huidige ontwerpkenmerken van de kunstwerken niet geschikt voor een efficiënt en effectief beheer en verdeling van irrigatiewater. Dit betekent een gebrek aan controle over het hoofd inlaatwerk, hydraulisch inconsistente combinaties van kunstwerken voor controle van de stroming, directe wateronttrekking langs de hoofd kanalen, open kunstwerken voor de wateruitlaat, onvoldoende afvoercapaciteit van kanalen en uitlaten en onvoldoende bescherming van de hoofd inlaatwerken tegen aanslibbing. Ook wezen een aantal studies op overschatting van de beschikbaarheid van oppervlaktewater en onderschatting van kwel en percolatie als oorzaken van de lage verhouding tussen geïrrigeerd gebied en irrigatie verzorgingsgebied.

De overheid van haar kant heeft in verschillende irrigatiesystemen renovatie en verbeteringsprojecten uitgevoerd. Ondanks aanzienlijke herstel en verbetering inspanningen, wordt ongeveer 30-40% van de irrigatie verzorgingsgebieden van het land niet geïrrigeerd. De veranderingen in het weerpatroon, de beschikbaarheid van water, gewassen, irrigatiewater en voorwaarden aan de dienstverlening, alsmede kunstwerken in irrigatiesystemen resulteren in de nodige problemen om de totale verzorgingsgebieden te irrigeren.

Modernisering van irrigatiesystemen wordt nu gezien als een proces van technische en bestuurlijke verbetering om de irrigatie dienstverlening aan de boeren op een hoger niveau te brengen. Deze inmiddels op grote schaal aanvaarde definitie van modernisering onderscheidt het van renovatie, waarbij gewoon het fysieke ontwerp van het originele systeem wordt hersteld. In het licht van de toenemende competitie om water en land en de toenemende kosten voor de ontwikkeling van nieuwe irrigatiesystemen, stellen experts dat het functioneren van bestaande kanaal irrigatiesystemen op het gebied van water en gewasproductie moeten verbeteren om de kans op het voldoen aan de wereldwijde behoefte aan voedsel in de nabije toekomst te verhogen. Zij zijn van mening dat modernisering van irrigatiesystemen een strategische optie is om het functioneren van kanaal irrigatiesystemen te verbeteren. Omdat veel irrigatiesystemen verouderd zijn, weerpatronen en hydraulische eigenschappen van waterbronnen veranderd en de wens van boeren om gewassen te intensiveren is toegenomen, zijn de fysieke structuur van bestaande irrigatiesystemen en de bijbehorende eisen aan het beheer van de kanalen minder geschikt om aan de huidige eisen voor irrigatie dienstverlening te voldoen. Zo zijn opwaardering van kunstwerken en het beheer van systemen noodzakelijke activiteiten geworden.

De algemene doelstelling van dit onderzoek was om technisch haalbare en hydrologisch solide moderniseringsplannen voor de nationale irrigatiesystemen in de Filippijnen te identificeren. Om deze doelstelling te bereiken, zijn de volgende studies uitgevoerd en de bevindingen geanalyseerd: (1) analyse van de aard van de vorige renovatie of modernisering inspanningen en de gevolgen daarvan in termen van het verschil tussen de werkelijk geïrrigeerde oppervlakte en de ontworpen gebieden; (2) de beoordeling van de effectiviteit van de kunstwerken en het beheer van irrigatiesystemen om de water hoeveelheden op basis van het ontwerp, de duur en de frequentie van irrigatiewater te leveren of een gewijzigde vraag naar irrigatiewater aan de uitlaten naar de secundaire kanalen te leveren; (3) beoordeling van de geldigheid van de ontwerp waarden voor betrouwbare watervoorziening, de waterbehoefte van de gewassen, kwel en percolatie; (4) identificeren van opties voor modernisering van realistische NIS verzorgingsgebieden; (5) onderzoek naar de mogelijkheden van ondiepe grondwater irrigatie met putten in het deel van de NIS verzorgingsgebieden die niet volledig kunnen worden geïrrigeerd met oppervlaktewater zoals ontworpen; (6) de identificatie van mogelijkheden voor het integreren van samengesteld gebruik van grondwater en oppervlaktewater bij het formuleren van een moderniseringsplan voor NIS.

Drie NIS die liggen in de regio IV-A van het land zijn geselecteerd voor de voorbeeld studie, te weten: het Agos Rivier Irrigatie Systeem (RIS), Balanac RIS en Sta. Maria RIS. De selectie was gebaseerd op een classificatie van de toestand van de fysieke voorzieningen als gevolg van eerdere renovatie werkzaamheden, derhalve, gereedheid voor overdracht van het irrigatiewater beheer (IMT). Bij een dergelijke indeling worden Balanac RIS en

Sta. Maria RIS geacht na kleine reparaties van een aantal van hun fysieke componenten klaar te zijn voor IMT. Terwijl voor Agos RIS grote herstelwerkzaamheden noodzakelijk worden geacht voordat het geschikt kan zijn voor exploitatie, onderhoud en beheer door boeren.

De gegevens van gedurende 1980-2012 gerenoveerde, gerestaureerde en nieuw aangelegde verzorgingsgebieden ten opzichte van de cumulatieve verzorgingsgebieden en werkelijk geïrrigeerde gebieden in het hele land zijn geanalyseerd om de effecten van de renovatie met betrekking tot het handhaven van ontwikkelde irrigatie verzorgingsgebieden en het verkleinen van het verschil tussen de verzorgingsgebieden en de werkelijk geïrrigeerde gebieden te bepalen. Ecn overeenkomstige trendanalyse is voor elk van de drie systemen uitgevoerd op basis van de gegevens over 1990-2012. De aard van de renovatie werkzaamheden aan de fysieke componenten die onderdeel waren van de renovatie projecten die sinds halverwege de jaren 1990 voor elk van de systemen zijn uitgevoerd zijn onderzocht. De belangrijkste personen op verschillende niveaus in de NIA kantoren en betrokken verenigingen van watergebruikers (WUA) zijn geïnterviewd om informatie en inzichten te verzamelen over het planning en realisatie proces van de renovatie projecten.

Van 1980 tot 2012 is het totale verzorgingsgebied van de NIS in het land met bijna 319.000 ha toegenomen. In dezelfde periode, bedroegen de nieuwe verzorgingsgebieden door de aanleg van nieuwe NIS en het uitbreiden van de verzorgingsgebieden van bestaande systemen ongeveer 415.000 ha. Een totale oppervlakte van ongeveer 3,5 miljoen ha NIS verzorgingsgebieden zijn tijdens deze periode gerenoveerd. Dit houdt in dat de NIS-verzorgingsgebieden in het land sinds 1980 eens in de 6-7 jaar zijn gerenoveerd. Gemiddeld, was voor elk decennium het gerenoveerde NIS gebied ongeveer 1,6 maal het verzorgingsgebied. Intussen was het gemiddelde jaarlijkse tempo van de renovatie voor de systemen in het kader van dit onderzoek ongeveer 75% van het verbeterde verzorgingsgebied (FUSA) van Balanac RIS en 40% van FUSA van Agos RIS en Sta. Maria RIS. De drie systemen zijn vaker gerenoveerd dan de andere systemen op nationaal niveau.

Op nationaal niveau, waren de renovatie projecten niet in staat om het irrigatie verzorgingsgebied te handhaven, noch om het verschil tussen het irrigatie verzorgingsgebied en het daadwerkelijk geïrrigeerde areaal weg te werken. In het geval van de drie systemen, konden door de renovatie, op zijn best, de percentages van de FUSA daadwerkelijk geïrrigeerd gebied sinds 2000 worden gehandhaafd. De omvang van de renovatie gaf het tempo van de verslechtering van de irrigatiesystemen in het land weer. Deze bevindingen wijzen op een aantal vragen met betrekking tot de efficiëntie van de planning en uitvoering van de kanaal irrigatieprojecten. Het doet twijfels rijzen over de effectiviteit van de renovatie.

De resultaten van de analyse van de onderliggende gegevens van de renovatie projecten laten zien dat de renovatie in het algemeen was gericht op het herstellen van de oorspronkelijke fysieke structuur. Bekleding van kanaal oevers met beton was de meest voorkomende activiteit en waarin ook het meest geïnvesteerd is, goed voor ongeveer 50-60% van de totale kosten van renovatie voor elke NIS. Hoofd inlaatwerken, wegen en kunstwerken behoorden in de drie NIS tot de volgende twee soorten projecten waarin het meest geïnvesteerd is. Ze betroffen elk minder dan 10% van de totale investering voor de respectievelijke systemen, met uitzondering van de dam van Balanac RIS en de weg in Sta. Maria RIS, die respectievelijk goed waren voor 22 en 15%.

Bij het toegepaste planning en het ontwerp proces ontbrak het opnieuw valideren van ontwerpcriteria en de veronderstellingen die waren gebruikt bij de oorspronkelijke ontwerpen van de systemen. Ook is er geen profijt geweest van diagnostische onderzoek studies. Er zijn geen technische evaluaties van de effectiviteit van de watervoorziening bij de voorgestelde projecten uitgevoerd. Bij het renovatie proces ontbrak een component om het effect ervan te evalueren.

Een combinatie van een logisch ontwerp kader door Ankum (2001), diagnose-instrumenten van het kartering systeem en methoden voor het beheer van kanalen (MASSCOTE), afvoer waterhoogte relaties en het hydraulische flexibiliteit concept zijn toegepast om het functioneren van Balanac RIS en Sta. Maria RIS te analyseren en mogelijkheden voor verbetering van de systemen the identificeren. Het logisch ontwerp kader onderzoekt de samenhang van de ontwerp filosofie, de algemene doelstellingen van een systeem, de doelstellingen van het beheer van het systeem, ontwerp configuratie van de kunstwerken en methoden voor het controleren van de stroming. Voor een set van de voorgestane ontwerpfilosofie en doelstellingen voor het systeem, bepaalde ontwerp parameters, systeemconfiguratie en combinaties van kunstwerken voor controle van de stroming passen goed bij elkaar of zijn misschien de enige logische mogelijkheid. Intussen zijn de diagnose-instrumenten van MASSCOTE, die bestaan uit een snelle beoordelingsprocedure (RAP) en de beoordeling van de fysieke capaciteit en het hydraulische gedrag (gevoeligheid en verstoringen) van irrigatiesystemen, gericht op systematische en uitgebreide evaluatie van het functioneren van grootschalige kanaal irrigatie systemen om het beheer van kanalen te moderniseren. Zij beschouwen de waterbalans, interne processen en mechanismen (hardware, operationele procedures, het beheer en de instellingen) van water voorziening op verschillende kanaal niveaus, geïrrigeerd areaal en de productie van gewassen in het diagnosticeren van het functioneren van het systeem.

Een overzicht van de algemene ontwerp overwegingen en richtlijnen voor kanaal irrigatie van de NIA en interviews met de staf van de irrigatie dienst betreffende de geschiedenis van de ontwikkeling van de systemen met inbegrip van vorige renovaties, veranderingen in het ontwerp van de systemen en het beheer zijn uitgevoerd. De ontwerpfilosofieën, systeemdoelstellingen, ontwerpconfiguraties van de kunstwerken en methoden voor controle van de stroming zijn ontleend aan deze verzamelde informatie en zijn geclassificeerd volgens de parameters van het logisch ontwerp kader.

De RAP werkbladen zijn gebruikt als een set van vragenlijsten voor het verzamelen van de informatie voor diagnose van de systemen. De informatie is verkregen uit documenten betreffende de systemen, van de meteorologische dienst alsmede via 'wandelingen door de systemen' en interviews met irrigatie opzichters, de operationele staf en verenigingen van watergebruikers. De 'wandelingen door de systemen' van de dam tot aan de laatste uitlaat van de hoofd kanalen en grote distributie kanalen van elk NIS zijn uitgevoerd in het kader van de RAP voor het verkrijgen van een indruk van de huidige staat van de kunstwerken en de werking van de systemen. De waarden van de RAP indicatoren betreffende het functioneren van de systemen zijn berekend uit de RAP computer spreadsheet gegevens.

De toepasbaarheid van de watergerelateerde externe RAP indicatoren en gevoeligheid evaluatie van de kunstwerken als diagnostische instrumenten was beperkt door het gebrek aan stromings gegevens en de onmogelijkheid om veld experimenten uit te voeren ten gevolge van de strakke rotatie irrigatie schema's en laissez-faire directe uitlaten. Daarom is een pragmatische aanpak toegepast door voor de analyses gebruik te maken van afvoer waterhoogte relaties, hydraulische flexibiliteit concept, de bevindingen tijdens de 'wandelingen door de systemen' en interviews.

Het resultaat van de indeling van de systeem profiel gegevens op basis van het logische ontwerp kader toonde aan dat het oorspronkelijke Balanac RIS systeem is ontworpen voor 'productieve irrigatie' tijdens het droge seizoen op basis van 'gelijkwaardige wateraanvoer per hectare'. De operationele doelstelling was 'opgelegde toewijzing' aan de tertiaire eenheden of irrigatie voorziening leveringspunten met 'instelbare aanvoer' en door 'instelbare aanvoer' door het hoofdsysteem of grote transport kanalen en het toepassen van bovenstroomse controle. Het was uitgerust met verstelbare verticale schuiven voor het regelen van afvoeren door de uitlaten en bovenstroomse waterstanden. Het oorspronkelijke ontwerp van Sta. Maria RIS had diezelfde filosofieën, de algemene doelstellingen van het systeem en de operationele doelstellingen, behalve in het geval van het ontwerp irrigatie seizoen. De ontwerp water leveringscapaciteit van 0,8 l/s/ha is zeer laag en suggereert logischer wijze dat het ontwerp 'irrigatie seizoen' het natte seizoen was. Met andere woorden, het systeem was ontworpen om de regenval aan te vullen en kon tijdens het droge seizoen niet het hele verzorgingsgebied irrigeren.

Er bestond een logische samenhang tussen de ontwerp filosofie of de algemene doelstellingen van de systemen, operationele doelstellingen en kunstwerken in de kanalen van het oorspronkelijke ontwerp van Balanac RIS en Sta. Maria RIS. Echter, de overgang naar 'gesplitste' stroming en 'proportionele controle' zoals gemanifesteerd door de eendenbek stuwen op de grote splitsingspunten van Balanac RIS was niet in overeenstemming met de ongewijzigde algemene doelstellingen van het systeem. In tegenstelling daarmee, is bij de vermindering van het verzorgingsgebied van Sta. Maria RIS en de overgang naar het droge seizoen als het belangrijkste irrigatie seizoen de samenhang tussen de doelstellingen gehandhaafd. Verder is de toevoeging van open directe uitlaten langs de hoofd kanalen van beide systemen niet in overeenstemming met de operationele doelstellingen van de systemen.

In het algemeen waren de berekende waarden van de primaire interne RAP indicatoren voor de twee systemen laag, variërend 0-2 (0 en 4 geven respectievelijk de minst en meest wenselijke waarde aan). De enige uitzonderingen waren in het geval van looptijd van een verandering in het debiet door de hoofd kanalen in beide systemen en in het geval van controle over de verdeelwerken langs de hoofd kanalen van Sta. Maria RIS. De resultaten van de RAP die is uitgevoerd in Balanac RIS ondersteunen de hypothese van starre water verdeling zoals afgeleid uit de logische ontwerp kader analyse. De waarden van de interne RAP indicatoren voor waterlevering en besturing en beheer van de kunstwerken waren laag voor de twee systemen, vooral door het gebrek aan functionele kunstwerken voor de controle van de stroming en de aanwezigheid van directe open uitlaten langs de hoofd kanalen en andere hoofd transport kanalen.

De capaciteit van de kunstwerken van Balanac RIS en Sta. Maria RIS om hun beoogde functies te realiseren is afgenomen als gevolg van schade, defecten, slecht functioneren, ontbrekende onderdelen en afwijkingen van de randvoorwaarden voor een goede werking door de tijd heen. Reparaties of vervangingen zijn nodig voor niet-functionele kunstwerken en onderhoudswerkzaamheden om de hoofd verdeelwerken op de dam te herstellen en de wateraanvoer condities naar de goten. In Balanac RIS en Sta. Maria RIS, waren de meest sprekende capaciteitsproblemen respectievelijk de verdeling capaciteit en de beperkte watertoevoer uit de rivieren. Mogelijkheden om de opslagcapaciteit van de Sta. Maria dammen te vergroten of de watertoevoer uit andere bronnen te verhogen moeten worden onderzocht.

De originele werdeelwerken bij de hoofd verdeelpunten van Balanac RIS en Sta. Maria RIS bestonden uit zowel verstelbare onderstroom afvoer en water niveau controles, die gevoelig zijn voor proportionele afleiding door de schuiven of het bereiken van dezelfde variaties in de afvoer bij de hoofd inlaat en uitlaat kunstwerken van de systemen. Echter, de benodigde frequente aanpassingen van de schuiven voor het handhaven van evenredige verdeling maakt de verstelbare onderstroom kunstwerken minder praktisch toepasbaar voor de dammen voor het afleiden van water uit de rivieren en voor variabele watertoevoer vanuit de rivieren. Het huidige gebrek aan functionele kunstwerken en de combinaties van overstroming en onderstroming bij hoofd verdeelpunten resulteren in niet proportionele variaties in de afvoer in de afleidingskanalen, de aansluitende kanalen en andere benedenstroomse uitlaten. De vele ongecontroleerde directe uitlaten dragen verder bij aan de inherente variaties in de stroming en verstoringen langs de hoofd kanalen.

Veldexperimenten en een grondige analyse van de ontwerp waarden en veronderstellingen betreffende percolatie en watervoorziening zijn uitgevoerd om de geldigheid en de juistheid van de geprojecteerde irrigatie verzorgingsgebieden te meten. Ook is de geldigheid van de aannamen beoordeeld in termen van het percentage van het geprogrammeerde irrigatiegebied het in feite geïrrigeerde gebied. Omdat kopieën van originele haalbaarheidsstudies en ontwerp documenten niet meer in de NIA kantoren aanwezig waren, zijn ontwerpers, hydrologen en de ingenieurs voor het beheer van de systemen van de NIA geïnterviewd over de procedures en methoden die door het agentschap zijn gebruikt bij het bepalen van de beschikbare waterhoeveelheden voor de irrigatie systemen. Verder zijn de procedures voor het ramen van de verzorgingsgebieden in haalbaarheidsstudies voor een steekproef van 20 irrigatieprojecten die in 2010-2014 zijn uitgevoerd geanalyseerd.

De percolatie testen met behulp van de 3-cilinder-methode en inundatie testen zijn op 19 plaatsen in de verzorgingsgebieden van Balanac RIS en Sta. Maria RIS uitgevoerd. Een gezamenlijke studie om de fysieke en hydraulische eigenschappen van de bodem ter plaatse van de experimenten te bepalen is met studenten uitgevoerd. De parameters die zijn geïdentificeerd omvatten de bodem textuur, bodemdichtheid, deeltjesdichtheid, porositeit en doorlatendheid. Ze werden respectievelijk bepaald met een hydrometer, oven drogen, volume verplaatsing en verval permeameter methoden. De resultaten van bodemtextuur analyses zijn gebruikt om de percolatie waarden voor elke grondsoort te bepalen zoals deze in de literatuur zijn vermeld en uiteindelijk de gemeten percolatie waarden te verifiëren.

De gemeten percolatie snelheden in Balanac RIS vertoonden een breed scala aan waarden, met een gemiddelde van 1-30 mm/dag. De ontwerp waarde van 2 mm/dag voor het systeem was slechts van toepassing in drie van de negen experiment locaties. De gemiddelde percolatie waarden waren groter dan 28 mm/dag in twee van de negen locaties. De respectievelijk gemiddelde percolatie in twee andere locaties verschilde significant indien de voorwaarden van de omringende bodem van de velden voor de experimenten veranderlijk was - van 5 mm/dag voor vaste bodem tot 22 mm/dag voor pas geploegde grond en van 12 mm/dag voor verzadigde grond tot minder dan 1 mm/dag voor onverzadigde grond. In Sta. Maria RIS varieerde de gemeten gemiddelde percolatie snelheden in de 10 locaties van 1-6 mm/dag. De ontwerp waarde van 1,4 mm/dag voor het systeem is waargenomen in vijf van de 10 experiment locaties. De gemiddelde percolatie waarden van ongeveer 6 mm/dag en 3-4 mm/dag werden verkregen in respectievelijk een locatie en vier experiment locaties.

De resultaten van de bodem textuur analyses toonden een aantal verschillen tussen de grondsoort in de locaties van de experimenten en die vermeld in de officiële bodemclassificatie kaarten, die variëren van klei tot keileem in Balanac RIS en klei in Sta. Maria RIS. Het meest opvallende verschil in Balanac RIS was de aanwezigheid van zand in drie experiment locaties. De percolatie snelheden verkregen op de locaties met zand waren

25-30 mm/dag, die veel hoger zijn dan de 2 mm/dag die gewoonlijk zijn gebruikt in de ontwerpen voor zavel leemgronden en 4 mm/dag voor zavel bodems. Keileem en leemgronden zijn ook gevonden in zes van de 10 experiment locaties in Sta. Maria RIS. Voor de vier locaties met overwegend kleibodems, hadden twee gemeten percolatie snelheden binnen de algemene ontwerp waarde van 1,25 mm/dag voor kleigronden. Ondertussen, had de meerderheid van de locaties met keileem percolatie snelheden die bijna twee keer zo groot waren als de conventionele ontwerp waarden van 1,75 mm/dag voor percolatie.

Hoewel de relatieve snelheden van de percolatie van de bodem- textuur indeling zijn waargenomen, verschilden de werkelijke percolatie waarden significant van de overeenkomstige waarden uit de literatuur en ontwerprichtlijnen. Andere factoren die de percolatie snelheid kunnen hebben beïnvloed waren de relatieve hoogte van de experiment locaties ten opzichte van de aangrenzende velden, opwellen van ondiep grondwater en ploegen dat de ploegzool losmaakt of breekt. Daarom zouden ten minste per tertiaire verzorgingsgebied metingen ter plaatse moeten worden gedaan.

De resultaten van de studie suggereren dat het potentieel voor het vergroten van de werkelijke oppervlakte geïrrigeerd gebied liggen in vergroting van watervoorziening of verhogen van de opslagcapaciteit van de systemen, het gebruik van kunstwerken voor controle van de stroming die geschikt zijn voor flexibele water verdeling schema's en strategische timing van irrigatie schema's over tertiaire gebieden aan te passen aan de heersende hydrologische regimes.

De ervaringen in de drie steekproef NIS gaven aan dat een meer systematische aanpak voor verbeteringen van irrigatiesysteem zou moeten worden gevolgd. Deze aanpak zou diagnostisch onderzoek van de fysieke structuur, beheer, onderhoud en water levering van irrigatiesystemen moeten omvatten; identificeren van beperkingen, mogelijkheden en opties voor verbetering; het vaststellen van voldoende water voor verbeteringsprojecten; inachtneming van de aanvaardbaarheid van verbetering opties voor de boeren; ontwikkeling van een blauwdruk van moderniseringsplannen voor individuele systemen en het tot stand brengen van een strategisch registratie en evaluatie systeem (M&E). Van de waarnemingen, analyses, bevindingen en aanbevelingen zoals beschreven in dit proefschrift wordt verwacht dat zij een solide basis voor dergelijke activiteiten creëren.

Annex Q. About the author

Mona Liza Fortunado Delos Reyes is since 2002 a University Researcher III at the University of the Philippines Los Baños (UPLB). In 1997 she obtained her BS Degree in Agricultural Engineering at the same university. This was in 2001 followed by her Master in Applied Science Degree (Agricultural Engineering) at the Massey University in New Zealand. Her MSc thesis was on the topic *Assessment of Water Availability for the Ororua River Catchment.*

Her work includes the preparation of research/project proposals, to perform research and extension activities on land and water resources development and enhancement, to write journal articles and to teach undergraduate courses (upon request by the Division and approval of an Authority to Teach by the Chancellor). In addition she performed several other duties, such as:

- project leader for the project *Modernization Strategy for National Irrigation Systems in the Philippines: linking design, operation and water supply*;
- consultant to the Philippine Institute for Development Studies (PIDS) for the project *Irrigation Governance - Irrigation rehabilitation component*;
- project Leader for the project *Assessing vulnerability to El Niño of Important Agricultural Commodities in the Different Agro-ecological Zones of the Philippines;*
- Project Leader for the project *Development and Initial Operation of an Irrigation Subsector Monitoring and Evaluation System*
- Irrigation Development Specialist for the Food and Agriculture Organization Representation in the Philippines (FAOR)

Publications

Delos Reyes MF, Schultz B, Prasad K (2016). Diagnostic assessment approach for formulating a modernization strategy for small-scale National irrigation systems in the Philippines. *Irrigation and Drainage,* (accepted for publication, on process).

Delos Reyes MF, David WP, Schultz B, Prasad K. 2015. Assessment of the process, nature and impact of rehabilitation for the development of modernization strategy for the national irrigation systems in the Philippines. *Irrigation and Drainage, 64: 464–478.*

Delos Reyes MF. 2014. Process, nature and impact of irrigation system rehabilitation. Philippines Institute for Policy Studies, Policy Note 2014-14.

David WP, Delos Reyes MF, Villano MG, Fajardo AL. 2012. Faulty Design Parameters and Criteria of Farm Water Requirements Result in Poor Performance of Canal Irrigation Systems in Ilocos Norte, Philippines. Philippine Agricultural Scientist, 95(2): 199-208.

David WP, Delos Reyes MF, Villano MG, Fajardo AL. 2012. Design Shortcomings of the Headwork and Water Distribution and Control Facilities of Canal Irrigation Systems in Ilocos Norte, Philippines. Philippine Agricultural Scientist, 95(1): 64-78.

Delos Reyes MF, David WP 2009. The Effect of El Niño on Rice Production in the Philippines. Philippine Agricultural Scientist, 92(2): 170-185.

David WP, Delos Reyes MF, Millare KL, Fajardo AL. 2007. Relative Vulnerabilities of the Different Provinces of the Philippines to El Niño-induced Drought of Various Magnitude. Philippine Agricultural Scientist, 90(4): 315-330.

Printed and bound by CPI Group (UK) Ltd, Croydon, CR0 4YY
18/10/2024
01776210-0003